# Smart Cities and Positive Energy Districts

# Smart Cities and Positive Energy Districts: Urban Perspectives in 2020

Editor

**Paola Clerici Maestosi**

MDPI • Basel • Beijing • Wuhan • Barcelona • Belgrade • Manchester • Tokyo • Cluj • Tianjin

*Editor*
Paola Clerici Maestosi
ENEA – TERIN
Smart Urban Network
Bologna, Italy

*Editorial Office*
MDPI
St. Alban-Anlage 66
4052 Basel, Switzerland

This is a reprint of articles from the Special Issue published online in the open access journal *Energies* (ISSN 1996-1073) (available at: https://www.mdpi.com/journal/energies/special_issues/EERA_JPSC_2020).

For citation purposes, cite each article independently as indicated on the article page online and as indicated below:

LastName, A.A.; LastName, B.B.; LastName, C.C. Article Title. *Journal Name* **Year**, *Volume Number*, Page Range.

**ISBN 978-3-0365-1188-7 (Hbk)**
**ISBN 978-3-0365-1189-4 (PDF)**

# Contents

# About the Editor

**Paola Clerici Maestosi** Dr., Phd, is a senior scientist at ENEA Energy Technologies and Renewables Department, Smart Urban Network Division, from 2010. She has been a practitioner from 1994 to 2000 in University of Tor Vergata, Real Estate Technical Unit, managing the complexity of construction and in-exercise activities at building and district level. Due to this extensive experience and to PhD she teaches as contract professor Building and Real Estate Management in Sapienza, Faculty of Architecture, Rome, from 2000 to 2007. In 2003 support Rome Municipality, Metropolitan Projects Department with activities dealing with district dimension. From 2010 she is working in ENEA as researcher in the area of Smart Cities and Positive Energy District (2018) providing expertise in development of smart and sustainable cities and energy communities, energy shift, energy landscape design and energy potential mapping. For well-known attitude in managing technical complexity, multi and transdisciplinary cooperation and multi-actors involvement she is the ENEA TERIN SEN delegate in EERA Joint Programme on Smart Cities (since 2010), then national delegate in Joint Programming Intitiative Urban Europe since 2012 (vice-chair from 2017), national referent for EERA IWG SET Plan Action 3.2 and, last but not least, National Scietific Expert (MUR Ministry of University and Research) for DUT Driving Urban Transition Partnership (2020). In the framework of EERA Joint Programme on Smart Cities she support the creation of the EERA Joint Programme on Smart Cities Special Issues Serie.

**EERA Joint Program on Smart Cities**

Since its creation, in 2010, EERA Joint Program on Smart Cities has accumulated important knowledge on specific topics of the programmes carried out. This includes Smart Cities and Positive Energy Districts, not only crucial topic tackled by EERA Joint Programme on Smart Cities Workplan but also by H2020, now Horizon Europe, and/or national call, focusing on innovative solutions based on interdisciplinary approaches, which are needed to face the highly complex structure of transition towards sustainable urban areas. Thanks to most prominent national and international RD&I programme, a lot of case studies, best practice and success case have been finalized, are ongoing, alternatively are in a planning stage. According to this a Special Issues Serie has been promoted to support publishing of the most promising research and innovation projects which EERA JPSC partners set up in the recent year so to drive the attention to the fact that EERA JPSC is one of the strong voice for the research area in Europe capable to highlight different points of view and solutions.
EERA Special Issues Serie already published:
1—2018: European Pathways for the Smart Cities to come;
2—2019: Tools, technologies and systems integration for the Smart and Sustainable Cities to come
with the ambition to publish the most promising research and innovation projects which EERA JPSC partners set up in the framework of H2020 Programme so to drive the attention to the fact that EERA JPSC is one of the strong voice for the research area in Europe capable to highlight different points of view and solutions. The EERA special issues serie scopes have been: 1—2018: European Pathways for the Smart Cities to come; 2—2019: Tools, technologies and systems integration for the Smart and Sustainable Cities to com.

 MDPI

Editorial

# Smart Cities and Positive Energy Districts: Urban Perspectives in 2020

**Paola Clerici Maestosi**

ENEA—Italian National Agency for New Technologies, Energy and Sustainable Economic Development, Department Energy Technologies and Renewables TERIN-SEN Smart Energy Networks, 40127 Bologna, Italy; paola.clerici@enea.it

**Abstract:** This Special Issue of *Energies* "Smart Cities and Positive Energy Districts: Urban Perspectives in 2020" introduce contemporary research on Smart Cities and on Positive Energy Districts. The topic highlights the variety of research within this field: from the analysis of 61 Positive Energy Districts cases to papers illustrating the Positive Energy Districts dimension or Smart Energy Communities supporting low carbon energy transition as well as selected Smart Cities Experiences. The focus is thus given on how RD&I stakeholders and Municipalities are facing sustainable urban development challenges. An overview of Horizon Europe RD&I program on sustainable urban areas is provided.

**Keywords:** Positive Energy District; sustainable urban areas; transition

**Citation:** Clerici Maestosi, P. Smart Cities and Positive Energy Districts: Urban Perspectives in 2020. *Energies* **2021**, *14*, 2351. https://doi.org/10.3390/en14092351

Received: 26 December 2020
Accepted: 26 January 2021
Published: 21 April 2021



## 1. Introduction

The United Nations Sustainable Development Goals (SDGs) are the global political agenda that addresses the most pressing social, economic and environmental challenges in an integrated way. The EU is committed to implementing the Sustainable Development Goals in all policies, including Research and Innovation (R&I) with the aim of modernizing the European economy and society and thus achieving a sustainable future. The EU Framework Program for Research and Innovation is one of the main vehicles promoting the transition to a sustainable future with the aim of becoming the first climate-neutral continent by 2050. All H2020 investments concerned at least one of the 17 SDGs, channeling 84% of investments on these.

While knowing that all the SDGs are equally important to achieve the goal of a sustainable future and that pathways to sustainability varies greatly, SDG 11 Sustainable Cities and Communities was funded most of all in the H2020 Framework Program of EU R&I with a relatively high share (43% to 58%) of total investment.

Thus, Europe capitalizes on over 30 years of investment in transnational EU Research and Innovation (R&I) on sustainable development. Within this timeline urban areas have been engaged in a multitude of initiatives aimed to boost sustainability dimension at urban scale, upgrading urban infrastructure and services, with a view to creating better environmental, social and economic conditions and enhancing cities' attractiveness and competitiveness. Reflecting these developments new "categories of cities" have entered the policy discourse but what still stands is that Horizon 2020 invested 43% of its budget (EUR 19.5 billion) on SDG 11 Sustainable Cities and Communities, mainly supporting Smart Cities Paradigm. In coming Horizon Europe the broad conceptualization of sustainable urbanization will stimulate cities, urban areas, and neighborhoods with an effective Research Development and Innovation program promoting 100 Positive Energy Districts and 100 Climate Neutral Cities within 2025. In this framework a reflection on the past and future of urban sustainable development is important to figure out next steps to support urban transition at the best. Thus EERA Joint Program on Smart Cities, a dedicated research network with own research program on Smart Cities and Positive Energy District, supported a special issues serie:

1. 2018-European Pathways for the Smart Cities to come, DOI: 10.13128/Techne-2356;
2. 2019-Tools, technologies and system integration for the Smart and Sustainable Cities to come, DOI: 10.5278/ijsepm.3515

Knowing that Horizon Europe Framework Program will deeply innovate Research, Development and Innovation for sustainable urban areas, the Scientific Board for EERA Joint Program on Smart City Special Issues Series where the author of this editorial acts as Coordinator, outlined topics of interest for third special issue:

- Smart Cities experiences
- Smart Energy Communities supporting the low carbon energy transition
- Positive Energy District

Moreover, as several international activities on sustainable urbanization are ongoing, the aims of this third special issue is to provide a clear reference framework about Horizon Europe Funding to boost transition to sustainable urban areas, which is highlighted by perspective paper [1] by Paola Clerici Maestosi guest editor of this special issue and Maria Beatrice Andreucci and Paolo Civiero as co-authors.

## 2. Published Papers Highlights

This Editorial article provides a summary of the Special Issue of *Energies*, covering the published papers [1–9] which address several of the topics mentioned in the Introduction. Table 1 identifies the most relevant topics in each published paper.

**Table 1.** Topic covered each publication.

| Topic | Publication | | | | | | | | | |
|---|---|---|---|---|---|---|---|---|---|---|
| | [1] | [2] | [3] | [4] | [5] | [6] | [7] | [8] | [9] | **TOT.** |
| Smart Cities experience | | X | X | | X | | X | | X | 5 |
| Smart Energy Communities supporting low energy transition | | X | | | X | X | | | X | 4 |
| Positive Energy District | X | | X | X | | X | | X | | 5 |

As shown in Table 1, most of the publications focus on Positive Energy Districts (5) while some of these (2) include someway contents deriving from Smart Cities experiences. Then we have five papers on Smart Cities experience and among these, two reflect on the Positive Energy District topic. This is clear evidence that a strong link exists among these two approaches which are able to boost sustainable urban transition. Finally four papers focus on Smart Energy Communities supporting the low carbon energy transition.

The nine articles have been selected after a peer review process and we are thankful to all the (32) forty-two authors from several countries (in alphabetic order: Austria, Denmark, Italy, Netherland, Poland, and Portugal) for their contribution to the Special Issue.

In the work [1], Justyna Patalas-Maliszewska and Hanna Łosyk present the results of an analysis about a Sustainable Urban Development Card for Zielona Góra. This Polish city which, in 2019, had the largest fleet of electric buses represent the case study of the research based on the data received from the Municipal Department of Transport of Zielona Góra. According to the data the values of SDGs were calculated for electric busses and then compared with conventional diesel busses. Moreover, the results of a questionnaire highlights an improved satisfaction of public transport passengers as well as an increased awareness in Sustainable Urban Development Goals. The conclusions of the article drive in the direction that the proposed Sustainable Urban Development Card is an innovative approach to monitoring and analyzing low carbon energy public transport and contributes to the aim of zero local emission which is one of the aims of energy communities.

The article [2] by Aleksandra Lewandowska et al. analyses and assesses the level at which renewable energy facilities are being implemented or developed in the urban space of cities in Poland as a pillar of the implementation of the smart city concept. Thanks to a multistage research procedure the authors stated that renewable energy installations already exist in many cities in Poland and are highly increasingly becoming important features in urban areas. Concerning renewable energy installation the research study demonstrates that the most popular is photovoltaic in buildings or as autonomous power systems. Among these small-scale RES installations are spreading fast, meanwhile large-scale ones are the result of the modernization of municipal sewage treatment plants, municipal heat and power plants and waste processing sites, where biogas and biomass are used in energy production. The authors conclusions are that identified place-based facilities prove that the smart city idea is increasing in popularity in Polish cities and it will represent an opportunity for cities, including post-socialist ones, to create a modern, functional, and environmentally friendly city (a smart city), and thus to build international market competitiveness.

Contributions [3] proposed by Nienke Maas et al. refers to energy flexibility aspects of Positive Energy Districts, addressing implementation conditions for energy flexibility technologies that facilitate cities to engender the expected impact and ensure replication and mainstreaming. The article capitalizes on the experience of EU H2020 Smart Cities and Communities project POCITYF, starting from the EU H2020 RUGGEDISED innovation and implementation framework for smart city technology, knowing that this framework is not concentrated on energy flexibility technologies alone. Thus in this paper the RUGGEDISED framework has been applied as a starting point to evaluate the implementation condition of energy flexibility solutions within the POCITYF project. Main findings rely on the area of services, legal framework and end users involvement.

The development of city-driven urban laboratories is considered a priority for setting up and replication of Positive Energy Districts in the article [4] by Silvia Soutullo et al. which refers to EERA Joint Programme on Smart Cities activity module 2. This strategic module assesses the development of city-driven Positive Energy Districts Laboratories (PED Lab). One of the objectives of EERA JP on SC module 2 is to identify research testing platforms (provided by members of EERA JP on SC) across countries in Europe that can be used for developing and testing integrated solutions and should work as drivers for local communities, districts, and cities. Once the information of the infrastructures submitted by the members of the EERA JP SC has been compiled, statistical studies have been performed to identify the main characteristics of these 16 facilities.

Sobah Abbas Petersen and all. [5] describe Smiling Earth, a mobile app to increase citizens' awareness about their own carbon footprint, by integrating energy and transport-related data. The article explores the ways in which ICT could help raise awareness and educate and motivate citizens about their actions and consequences on the environment. The value of the Smiling Earth for individuals, cities, and communities are presented in the conclusion in terms of benefits to individuals, cities, and neighborhoods and benefits to citizens and communities of people, as well as the role of technologies from the different perspectives of an individual, a community or a city.

In [6], Dorota Chwieduk et al. evaluate the possible transition routes from the existing centralized energy systems in Polish cities to modern low emission distributed energy systems based on locally available energy sources, mainly solar energy. The evaluation of these possibilities is based first on the presentation of the current structure of energy grids and heating networks in Polish cities, then a basic review of energy consumption in the building sector is given with the emphasis on residential buildings. The paper deals with evaluation of the effectiveness of operation of central district heating systems and heat distribution systems, predicts the improvement in the effectiveness of the energy production, distribution and use, and analyses the possible integration of existing systems with distributed energy sources.

A new digital Geographic Information System (GIS) platform developed to quantify the energy savings obtained through the implementation of refurbishment measures in residential buildings, including solar thermal collectors and geothermal technologies, and assuming the postal district as the representative unit for the territory is presented in [7] by Emanuela Giancola et al. The new developed GIS platform based on the interaction of different technologies and available resources is a step forward in the field of built environment towards smart energy communities, supporting the low carbon energy transition. The article highlights the GIS methodology which has been applied to the Asturian region in Spain, qualified as a region in energy transition. The obtained results point to great energy mitigation potentials justifying the need for public authorities to develop policies aimed at reducing the energy demand of buildings, especially the heating loads. The generated maps could be helpful for retrofitting considerations for existing housing stock in the Principality of Asturias, a key action that is completed in combination with the use of solar thermal energy and geothermal resources towards sustainable development.

Silvia Bossi et al., in [8], proposes a solid understanding and consideration based on analysis of cities' strategies, experiences and project features which are the basis of developing and designing the PED program. JPI Urban Europe has been collecting information on projects towards sustainable urbanization and the energy transition across Europe. The collected cases are summarized in a PED Booklet whose update was recently published on the JPI Urban Europe website. Results presented in this paper provide insights from the analysis of 61 projects in Europe and offer recommendations for future PED developments. There is a broad variety of projects addressing energy transition. While only 29 of 61 projects are planning to implement PEDs, all of them provide valuable approaches and strategies regarding actions towards the energy transition in the urban context. Most of the projects classified as Towards PEDs seem to include the possibility to develop into PEDs in a further step. However, actual feasibility of many project designs is not yet accessible since a large share of the projects listed are in planning or early implementation stages, respectively. The PED Reference Framework with its focus on energy functions and overall sustainability goals provides an open framework; he results of the comparison between PED projects and Towards PEDs projects indicate that differences between the two groups are not categorical and the main issues and challenges of project implementation are similar. This, therefore, supports the assumption that projects with no self-declared PED ambition may be further developed to achieve PED status according to the PED Reference Framework. A further operationalization of the PED Reference Framework [5], which is currently under development, aims at a common PED definition that includes local differences and diverse approaches.

**Conflicts of Interest:** The author declares no conflict of interest.

## References

1. Maestosi, P.C.; Andreucci, M.B.; Civiero, P. Sustainable Urban Areas for 2030 in a post-Covid-19 Scenario: Focus on innovative research and funding frameworks to boost transition towards 100 Positive Energy Districts and 100 Climate-Neutral Cities. *Energies* **2021**, *14*, 216. [CrossRef]
2. Lewandowska, A.; Chodkowska-Miszczuk, J.; Rogatka, K.; Starczewski, T. Smart energy in a smart city: Utopia or reality? Evidence from Poland. *Energies* **2020**, *13*, 5795. [CrossRef]
3. Maas, N.; Georgiadou, V.; Roelofs, S.; Lopes, R.A.; Pronto, A.; Martins, J. Implementation Framework for Energy Flexibility Technologies in Alkmaar and Évora. *Energies* **2020**, *13*, 5811. [CrossRef]
4. Soutullo, S.; Aelenei, L.; Nielsen, P.S.; Ferrer, J.A.; GonÇalves, H. Testing platforms as drivers for positive energy living laboratories. *Energies* **2020**, *13*, 5621. [CrossRef]
5. Petersen, S.A.; Petersen, I.; Ahcin, P. Smiling Earth–Raising Awareness among Citizens' for Behaviour Change to Reduce Carbon Footprint. *Energies* **2020**, *13*, 5932. [CrossRef]

6. Chwieduk, D.; Bujalski, W.; Chwieduk, B. Possibilities of transition from centralised energy systems to distributed energy sources in large Polish cities. *Energies* **2020**, *13*, 6007. [CrossRef]
7. Soutullo, S.; Giancola, E.; Sánchez, M.N.; Ferrer, J.A.; García, D.; Súarez, M.J.; Prieto, J.I.; Antuña-Yudego, E.; Carús, J.L.; Fernández, M.Á.; et al. Methodology for quantifying the energy saving potentials combining building retrofitting, solar thermal energy and geothermal resources. *Energies* **2020**, *13*, 5970. [CrossRef]
8. Bossi, S.; Gollner, C.; Theierling, S. Towards 100 Positive Energy Districts in Europe: 3 preliminary data analysis of 61 European cases. *Energies* **2020**, *13*, 6083. [CrossRef]
9. Patalas-Maliszewska, J.; Łosyk, H. Analysis of the Development and Parameters of a Public Transport System which uses Low-carbon Energy–the Evidence from Poland. *Energies* **2020**, *13*, 5779. [CrossRef]

*Article*

# Towards 100 Positive Energy Districts in Europe: Preliminary Data Analysis of 61 European Cases

**Silvia Bossi [1,*], Christoph Gollner [2] and Sarah Theierling [2]**

[1] Italian National Agency for Technologies, Energy and Sustainable Economic Development, ENEA, CR Casaccia, 00123 Rome, Italy

[2] Austrian Research Promotion Agency, FFG, 1090 Vienna, Austria; Christoph.Gollner@ffg.at (C.G.); sarah.theierling@ffg.at (S.T.)

* Correspondence: silvia.bossi@enea.it

Received: 12 October 2020; Accepted: 16 November 2020; Published: 20 November 2020

**Abstract:** Positive Energy Districts and Neighborhoods (PEDs) are seen as a promising pathway towards sustainable urban areas. Several cities have already taken up such PED-related developments. To support such approaches, European countries joined forces to achieve 100 PEDs until 2025 through a comprehensive research and innovation program. A solid understanding and consideration of cities' strategies, experiences and project features serve as the basis for developing and designing the PED program. JPI Urban Europe has been collecting information on projects towards sustainable urbanization and the energy transition across Europe. The collected cases are summarized in a PED Booklet whose update was recently published on the JPI Urban Europe website. Results presented in this paper provide insights from the analysis of 61 projects in Europe and offer recommendations for future PED developments.

**Keywords:** Positive Energy District (PED); PED Booklet; SET-Plan

---

## 1. Introduction

Cities have a strong impact on greenhouse gas production and climate change, and several programs and initiatives are focused on new strategies to transform cities into climate neutral ones [1,2].

The Strategic Energy Technology (SET) Plan, adopted by the European Union in 2008 to establish an energy technology strategy in that region, is focused on 10 key action fields, of which Action 3.2 aims to support the planning, deployment and replication of 100 Positive Energy Districts by 2025 for sustainable urbanization [3].

The Program on Positive Energy Districts and Neighborhoods (PED Program) is conducted by JPI Urban Europe in cooperation with the SET Plan Action 3.2, and it involves stakeholders from city administrations, urban planning, industry, research, and civic society organizations [4]. This is seen as an important contribution to current policies on achieving the goals of reducing Europe's carbon footprint, managing the energy transition, and cities' ambitions towards sustainable urban development. A solid understanding and consideration of cities' experiences and strategies serves as the base of developing the program. This is why the PED program aims at a strong involvement of all stakeholder groups in the program's implementation.

A framework for PED has been defined as follows: "Positive Energy Districts are energy-efficient and energy-flexible urban areas or groups of connected buildings which produce net zero greenhouse gas emissions and actively manage an annual local or regional surplus production of renewable energy. They require integration of different systems and infrastructures and interaction between buildings, the users and the regional energy, mobility and ICT systems, while securing the energy supply and a good life for all in line with social, economic and environmental sustainability" [5]. The primary energy is the performance indicator of the PED definition and the balance includes both building operations and user-related energy consumptions. Participating PED program partners have agreed that this PED reference framework offers a common baseline across all countries while ensuring flexibility regarding local conditions for PEDs at the same time. Consequently, guidelines for identifying geographical and virtual boundaries of PEDs are still under discussion in collaboration with other programs (i.e., EERA Joint Program Smart Cities) with the ambition of making PEDs achievable for wide range of possible approaches. Examples of PED projects are already under development in Europe in connection with research activities [6–8].

A PED Program mapping of urban projects dealing with the energy transition and sustainability aims to provide an overview of projects, approaches and strategies towards Positive Energy Districts. A compilation of these current cases serves as a basis for knowledge exchange and identification of good practice. The collected cases are summarized in a PED Booklet, an update of which was recently published on the JPI Urban Europe website [9].

Of the 61 cases described in the Booklet, 29 declared a PED ambition. This paper analyses data collected in the PED Booklet to identify common features, strategies, challenges, and success factors of PED projects to guide stakeholders in this field.

For this analysis it must be considered that not all countries have equally contributed to the Booklet yet. The Booklet is based on voluntary contributions from individual PED-related projects and will be continuously updated. Therefore, the results do not allow any interpretation about the level of PED efforts across Europe in general yet.

## 2. Materials and Methods

For this study, the 61 cases described in the PED Booklet (later referred to as Booklet) were considered and analyzed. Project descriptions were collected using a template that was sent out to reference persons of each project. The template was organized into 4 sections:

- General information: this section included information about geographical position, contacts, size, building structure, land use, and financing. Some of these questions allowed multiple choice answers.
- Overview description of the project: this section provided with a text description with the possibility to add further project information and illustrations.
- Strategies: this part included multiple choice questions about goals/ambitions, expected impact, overall strategy of city municipality, factors included in implementation strategies, stakeholder involvement strategies, and typology of energy supply.
- Success factors, challenges, and barriers: text descriptions were collected on key learnings regarding supportive and inhibiting factors in the PED implementation process.

The first question in the "Strategies" section of the template was on goals and ambitions. Among the multiple choices, there was the option "Positive Energy". The cases were divided in two categories: projects that declared a PED ambition (PED, $N_P$ = 29) versus projects that did not declare a PED ambition but presented interesting features for the PED Program (Towards (To)-PED, $N_{TP}$ = 32).

The data were collected in an Excel file table and analyzed separately for PED and To-PED. Only features with less than 30% of n/a data were selected as reliable and worthy to be analyzed:

- Geographic distribution
- Implementation status
- Building structure
- Land use
- Energy typology
- Success factors, challenges, and barriers

## 3. Results

### 3.1. Geographic Distribution of PED and Towards PED Projects

The Booklet includes 61 cases in 19 different countries in Europe (Table 1). Of the 61 projects, 29 projects, located in 13 European countries, declared a PED ambition while the rest of the projects (To-PED) are located in 14 European countries.

**Table 1.** Geographic distribution of projects.

| Country | PED | To-PED | TOT |
|---|---|---|---|
| Norway | 8 | 1 | **9** |
| Finland | 4 | 3 | **7** |
| Sweden | 1 | 5 | **6** |
| The Netherlands | 3 | 3 | **6** |
| Italy | 3 | 5 | **8** |
| Germany | 0 | 4 | **4** |
| Denmark | 0 | 2 | **2** |
| Austria | 2 | 2 | **4** |
| France | 1 | 1 | **2** |
| Spain | 2 | 2 | **4** |
| Switzerland | 0 | 1 | **1** |
| Romania | 1 | 0 | **1** |
| Estonia | 1 | 0 | **1** |
| Ireland | 1 | 0 | **1** |
| Belgium | 0 | 1 | **1** |
| Greece | 0 | 1 | **1** |
| Hungary | 0 | 1 | **1** |
| Turkey | 1 | 0 | **1** |
| Portugal | 1 | 0 | **1** |
| **TOT** | **29** | **32** | **61** |

The highest number of projects of the Booklet (PED + To-PED) are located in Norway (9), Italy (8), Finland (7), Sweden (6), and The Netherlands (6).

In terms of PED projects, Norway contributed the most cases (8). On the contrary, only two PED/To-PED projects are located in Eastern and Southeastern Europe.

This geographical imbalance may be explained mainly with either of two factors: (1) policy priorities and implementation status of policies towards the energy transition vary between European countries; therefore, specific national programs on PEDs or PED-related matters have only been implemented in some countries. An example is represented by the ZEN Research Centre (https://fmezen.no/about-us/); nine pilot projects were spread over Norway with the aim of developing zero emission neighborhoods [10–12]. ZEN has public and industrial partners to guarantee a multidisciplinary approach that combines together researchers with building professionals, property developers, public authorities, energy companies, building owners, and users. (2) High support of national delegates for the PED Program in collecting information on PED projects in their countries.

### 3.2. Implementation Status

The template distributed to projects' representatives included a question related to the implementation status of their projects. Possible choices were: "Planned", "Implementation Stage", "Realized", "In Operation", and "n/a". Percentages of PED and Towards PED projects in different implementation phases are summarized in Figure 1. Most of PED projects (green) are in their implementation stage (69%); 24% are in planning stage and only a minority are realized (3%) or already in operation (3%). To-PED projects (in blue) are mainly in their implementation stage (44%), but also a considerable amount of them are already in operation (31%); 9% are realized but not yet in operation.

Only 13% of To-PED projects are in planning stage. As expected, PED projects are less mature than Towards PED projects; among them only one project is already in operation and only one is realized.

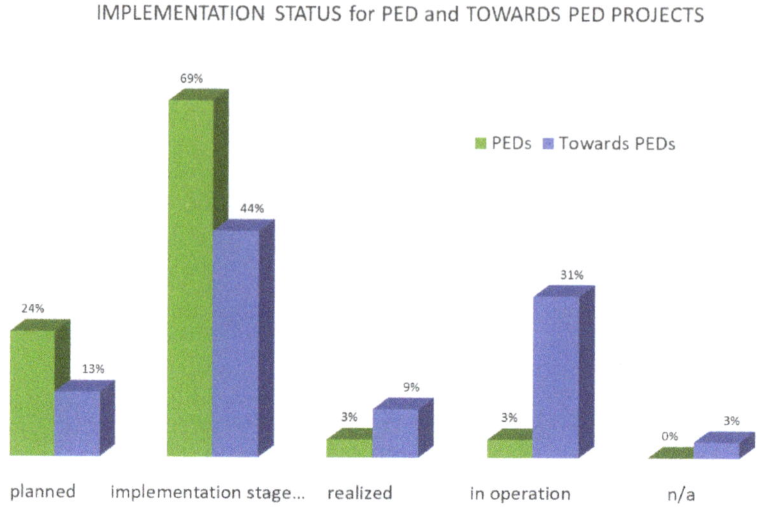

| | PED | | To- PED | |
|---|---|---|---|---|
| **Project Status** | **in Figures** | **in %** | **in Figures** | **in %** |
| planned | 7 | 24.14% | 4 | 12.50% |
| implementation stage | 20 | 68.97% | 14 | 43.75% |
| realized | 1 | 3.45% | 3 | 9.38% |
| in operation | 1 | 3.45% | 10 | 31.25% |
| Project status n/a | 0 | 0.00% | 1 | 3.13% |

**Figure 1.** Implementation status of Positive Energy District and Neighborhood (PED) versus Toward (To)-PED.

### *3.3. Building Structure*

Differences of projects' building structures were analyzed by dividing them into "Newly Built", "Existing Neighborhood" and "Mixed".

Projects with PED ambition mainly show a mixed typology (Figure 2); 66% of them combine newly built with existing neighborhoods, 28% of them are newly built districts, and only 7% are developed from existing neighborhoods.

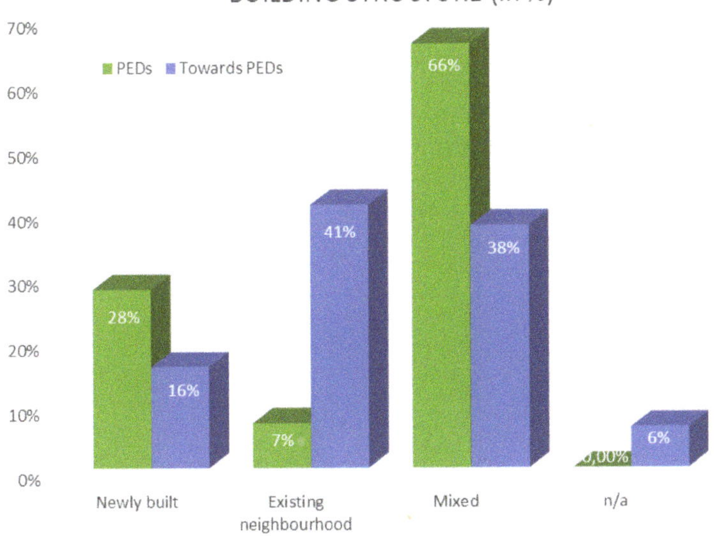


| | PED | | To- PED | |
|---|---|---|---|---|
| **Building structure** | in Figures | in % | in Figures | in % |
| Newly built | 8 | 27.59% | 5 | 15.63% |
| Existing neighbourhood | 2 | 6.90% | 13 | 40.63% |
| Mixed | 19 | 65.52% | 12 | 37.50% |
| Building structure n/a | 0 | 0.00% | 2 | 6.25% |

**Figure 2.** Building structure of PED versus To-PED.

On the contrary, To-PED projects are mainly based on existing neighborhoods (41%). A considerable percentage of them is based on a mixed typology (38%) and only 16% of Towards PED projects are newly built. A comparison among PED and Towards PED projects is shown in Figure 2.

PEDs results are in good agreement with the framework, which suggests the combination of new urban development areas with existing buildings.

### 3.4. Land Use

The Booklet includes information on land use of projects areas. Multiple choices were available for this topic: "Residential", "Office", "Industry", "Commercial", "Social", "Other", and "n/a".

Twenty-three PED and 17 To-PED projects specified a land use and results are summarized in Figure 3a. All the projects (100% of PED and 100% of To-PED) included residential use. In addition to residential use, 65% of PED projects included a social land use versus 29% of To-PED projects. Commercial activities were foreseen in 61% of PED and only in 24% of To-PED projects. More comparable percentages were observed for office spaces (61% of PED versus 53% of To-PED) and industry (22% of PED versus 18% of To-PED). In addition, 30% of PED and 47% of To-PED projects mentioned other typologies of land use (e.g., logistic hubs, natural facilities, waste water energy recovery systems, etc.).

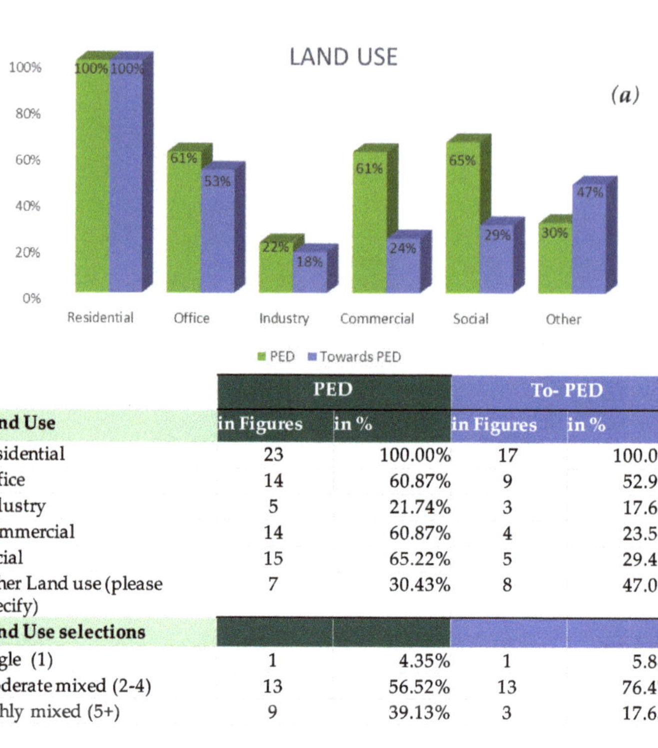


| | PED | | To- PED | |
|---|---|---|---|---|
| **Land Use** | in Figures | in % | in Figures | in % |
| Residential | 23 | 100.00% | 17 | 100.00% |
| Office | 14 | 60.87% | 9 | 52.94% |
| Industry | 5 | 21.74% | 3 | 17.65% |
| Commercial | 14 | 60.87% | 4 | 23.53% |
| Social | 15 | 65.22% | 5 | 29.41% |
| Other Land use (please specify) | 7 | 30.43% | 8 | 47.06% |
| **Land Use selections** | | | | |
| single (1) | 1 | 4.35% | 1 | 5.88% |
| moderate mixed (2-4) | 13 | 56.52% | 13 | 76.47% |
| highly mixed (5+) | 9 | 39.13% | 3 | 17.65% |

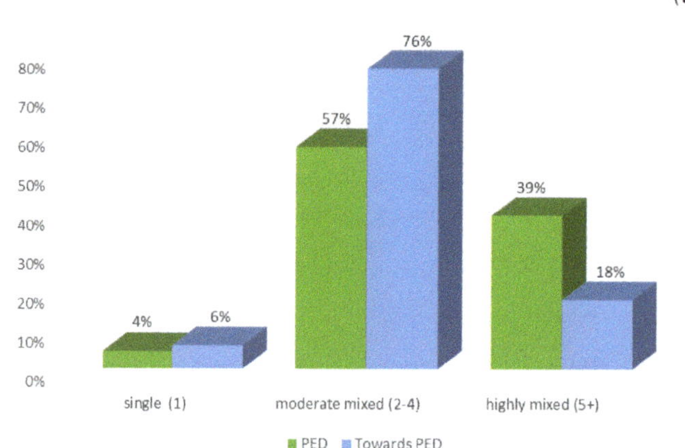


**Figure 3.** (**a**) Land use of PED versus To-PED; (**b**) land use selection of PEDs versus To-PED.

The number of different land use selections were counted for PED and Towards PED projects. The results reflect the degree of mixed use in the projects and were divided into three groups: "single use", "moderate mixed" (two to four land uses), and "highly mixed" (more than five land uses). Comparison results for PED and To-PED projects are summarized in Figure 3b. Only small percentages of PED and To-PED projects have a single land use. A high percentage of PED and To-PED projects have a moderate mixed land use, namely 57% and 76%, respectively. A significant amount of PED projects has a highly mixed use (39%), whereas the same is true in only 18% of Towards PED projects.

### 3.5. Energy Typology

The template used so far included a question on "Typology of Energy Supply" where project representatives could select one or more of the following options: "Solar thermal energy", "Geothermal energy", "District heating/local heating", "Heat pump system", "Industrial waste heat", "Photovoltaic", "Other", and "n/a".

Answers to this question where collected for PED and To-PED projects and the results in percentage of total projects with energy data available are shown in Figure 4a. Most of the projects included "district/local heating" in their energy strategy, with similar percentages for PED (75%) and To-PED projects (83%). Several projects included "heat pump system", with 71% of PED and 60% of To-PED projects. Among PED projects, 71% of them exploited photovoltaic energy, while only 30% of To-PED projects used PV. The share of PV use represents the main difference between the two project categories, followed by the use of geothermal energy—58% for PED versus 31% of To-PED. Half of PED use solar thermal energy versus 37% of To-PED; a similar result for PED and To-PED was observed for industrial waste heat (25% and 33%, respectively). Other typologies of energy supplies described by the PED projects included wind, biomass, waste digestion, and hydrothermal.

For each project, we evaluated the number of different typologies of energy supply combined together. Data obtained were classified in four different groups based on the number of energy sources used:

- Single (only one)
- Moderate mixed (two to three)
- Highly mixed (four to five)
- Very highly mixed (less than six)

The quantification of multiple energy sources integration for PED and To-PEDs is summarized in Figure 4b. Half of PED projects use "highly mixed" energy sources (4 or 5), 29% are "moderate mixed", 13% are "very highly mixed", and only 8% use just a single source of energy.

Most of To-PED projects are "moderate mixed"; 60% use two or three sources of energy, 30% are "highly mixed", 7% use only a single source, and only 3% use "very highly mixed" energy typologies.

Results suggest that PEDs combine more energy sources (four or five) than Towards PED projects (two or three) to satisfy the energy surplus required. PED projects have the aim of reaching an energy surplus production from renewable energies. Combining a high number of energies is not the aim for a PED but it seems to be a frequent approach of PED projects. PED preferred energy typologies are district and local heating, heat pump systems, photovoltaic, and geothermal energy.

### 3.6. Success Factors and Challenges of Projects Implementation

An analysis of the statements regarding success factors and challenges/barriers of PED and To-PED projects (Figures 5 and 6) shows the perceived importance of stakeholder involvement. In fact, involvement processes (urban stakeholders, citizens) can be seen as defining elements for the success or failure of a project. Political support (or lack of it), most likely connected with funding, is seen as another key aspect, especially for PED. Capacity-building among decision-makers can thus be seen as a highly important task. Not only funding, but the elaboration of feasible business models also ranks very high among key aspects of PED development.

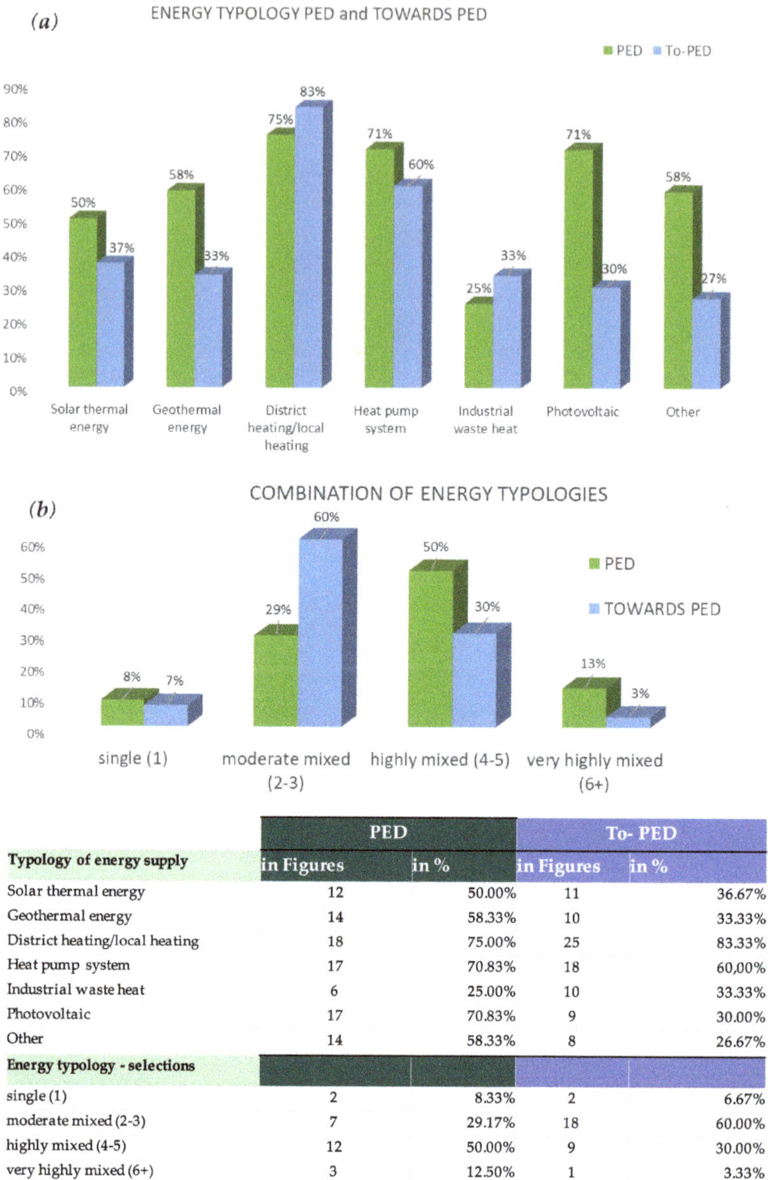


| | PED | | To- PED | |
|---|---|---|---|---|
| **Typology of energy supply** | **in Figures** | **in %** | **in Figures** | **in %** |
| Solar thermal energy | 12 | 50.00% | 11 | 36.67% |
| Geothermal energy | 14 | 58.33% | 10 | 33.33% |
| District heating/local heating | 18 | 75.00% | 25 | 83.33% |
| Heat pump system | 17 | 70.83% | 18 | 60,00% |
| Industrial waste heat | 6 | 25.00% | 10 | 33.33% |
| Photovoltaic | 17 | 70.83% | 9 | 30.00% |
| Other | 14 | 58.33% | 8 | 26.67% |
| **Energy typology - selections** | | | | |
| single (1) | 2 | 8.33% | 2 | 6.67% |
| moderate mixed (2-3) | 7 | 29.17% | 18 | 60.00% |
| highly mixed (4-5) | 12 | 50.00% | 9 | 30.00% |
| very highly mixed (6+) | 3 | 12.50% | 1 | 3.33% |

**Figure 4.** (**a**) Energy typology of PED versus To-PED; (**b**) energy typology selections of PED versus To-PED.

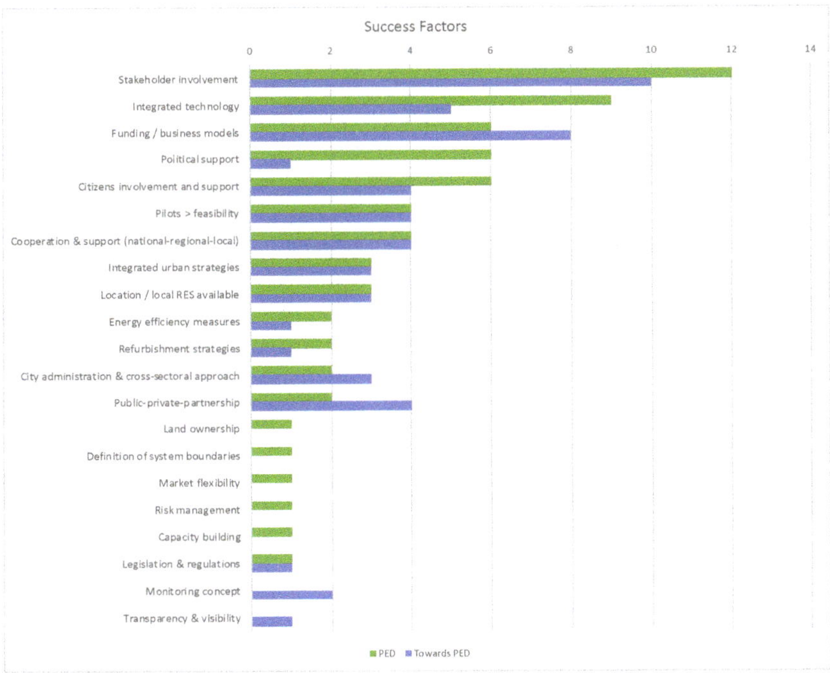


**Figure 5.** Success factors of implementation of PED and To-PED projects.

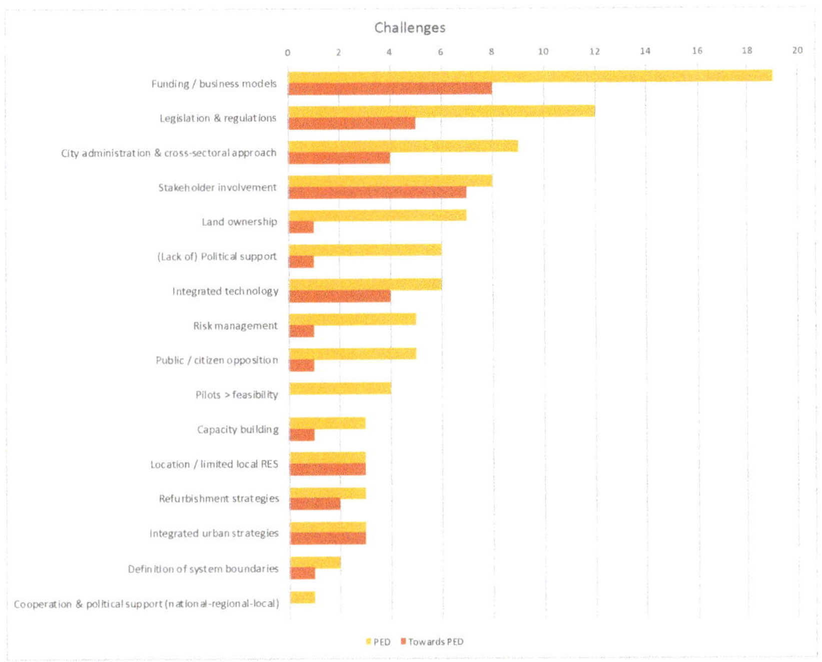


**Figure 6.** Challenges and barriers for implementation of PED and To-PED projects.

A big challenge for successful implementation processes are legislation and regulations—an in-depth analysis of the particular legal barriers (on either local, national, or European levels) therefore seems to be highly relevant in particular for PED. Within the +CityxChange project, possible strategies to enable regulatory mechanisms have been studied [13].

## 4. Discussion and Conclusions

The analysis presented in this report was performed on data included in the first update of the PED Booklet (February 2020). The aim of this study is to evaluate trends, features, successes, or critical factors that contribute to development of PED projects in Europe.

There is a broad variety of projects addressing energy transition. While only 29 of 61 projects are planning to implement PEDs, all of them provide valuable approaches and strategies regarding actions towards the energy transition in the urban context. Most of the projects classified as Towards PEDs seem to include the possibility to develop into PEDs in a further step. However, the actual feasibility of many project designs is not yet assessable since a large share of the projects listed are in planning or early implementation stages. The PED Reference Framework, with its focus on energy functions and overall sustainability goals, provides an open framework—the results of the comparison between PED projects and Towards PEDs projects indicate that differences between the two groups are not categorical and the main issues and challenges of project implementation are similar. This therefore supports the assumption that projects with no self-declared PED ambition may be further developed to achieve PED status according to the PED Reference Framework. A further operationalization of the PED Reference Framework [5], which is currently under development, aims at a common PED definition that includes local differences and diverse approaches.

For this analysis it must be considered that not all countries have equally contributed to the Booklet yet. The Booklet is based on voluntary contributions from individual PED-related projects. Therefore, the results do not allow any interpretation about the level of PED efforts across Europe in general.

Nevertheless, these preliminary results highlight the importance of national programs and strong R&I involvement in the replication of PED strategies, as for example in Norway where ZEN Research Center on Zero Emission and Neighborhoods in Smart Cities is helping the spreading of pilot projects in line with the PED framework. Future investigations on the impact of climatic zones on PED mainstreaming will be also considered.

Projects that declared a PED ambition are mainly in their planning or implementation stage (93% versus 57% of To-PED). This suggests that most of PED strategies are still to be tested and optimized and a regular update of the Booklet cases and data analysis is needed. Efforts will continue to extend the data base, improve data quality, and assess these projects to identify good practice as well as critical issues at further stages.

Most projects with PED ambitions (66%) combine newly built neighborhoods with existing neighborhoods. This result is in good agreement with the framework that suggests the combination of new urban development areas with existing buildings. Only 7% of PED are developed in an existing neighborhood. The high complexity of a PED project requires deep interventions to fit with the new paradigm and might inhibit ambitions to apply PED ambitions to existing structures. The need for dealing with complex ownership structures or technological limitations, for instance regarding PV-instalment, in connection with other aspects of revitalization, for example, building protections and preservation of cityscape, are just two factors that need to be mentioned. Regarding the implementation of a PED ambition, it does not seem to be a common strategy to count only on existing neighborhood. However, for most European cities, transformation of the existing building stock is a key aspect for achieving the energy transition. Therefore, exploring, integrating and applying PED solutions and approaches to the existing urban structure seems to be a crucial challenge in the long run. An 8-step process to develop a PED is proposed by the JRC report [14].

A significant amount of PED projects involves a highly mixed land use (39%) while only 18% of Towards PED projects do. Multifunctional urban structures can be seen as a basic requirement for successful PED implementation since they support efficient use of energy and more opportunities for energy flexibility. A high percentage of PED mentioned commercial and social use of the land. Multifunctionality is also a crucial requirement for embedding PED strategies in integrated urban planning strategies aiming at high quality of life and sustainable local development. This is in line with the framework definition that requires social and economic sustainability.

Another feature of PED is their highly mixed combination of energy typologies. Fifty percent of PED projects integrate four to five energy sources, whereas 60% of To-PED projects use two or three. This is a reasonable result given the fact that PED projects have the ambition to have an annual positive energy balance produced from renewable sources. To satisfy this ambitious aim, the combination of multiple energy typologies is key to exploit changing energies' availabilities, consumptions with different weather conditions, or with the alternation of day and night and seasons. Photovoltaics is used in 71% of PED projects versus 30% of To-PED projects. Photovoltaics is a mature and suitable technology for decentralized and renewable energy production. Its profitability is already reached in some European countries and increasing diffusion in building rooftop is expected [15]. This explains why a high percentage of PEDs rely on photovoltaics solar energy as local production of renewable energy.

Stakeholders, citizens involvement strategies, integrated technology, and political support are considered main success factors for PED development. A successful PED implementation requires collaboration with citizens and final users right from initial planning stages to avoid their opposition to the change of paradigm that a PED aims to do in terms of social, economic and energy sustainability. Political support is a necessary factor to activate national programs and new funding opportunities. Furthermore, the technical complexity of a PED is the reason why integrated technology was mentioned as a main feature to its success.

The main barrier for a PED is the access to adequate funding and business models (Figure 6). PEDs are more expensive than traditional projects and they require multiple sources of financing and advanced business models. Innovations and changes introduced by PED projects require ad hoc regulations and a dedicated legal framework. This explains why legislation and regulations, together with existing governance structures within city administrations and a lack of cross-sectoral approaches, are often mentioned as main reasons for inhibited implementation processes (Figure 6).

Considering the issues discussed, key areas of action to be further explored can be summarized as follows:

- Allowing for and testing of a diversity of approaches and strategies: exploring room for experimenting with and implementing of PED Labs; monitoring existing approaches the and elaboration of guides and tools
- Developing fitting governance structures and the legal framework: consideration of Functional Urban Areas (regional perspective); inter-departmental cooperation within city administrations and collaboration between city administrations, energy providers, real estate industry, civic society, and other stakeholders; re-consideration and adaptation of legislation and regulations
- Integration of PED strategies in holistic urban planning: mainstreaming of energy planning in urban planning strategies; connecting energy aspects with climate action; a high quality regard for functions and design
- Developing feasible business models and funding opportunities: creating awareness and political support for development and implementation of national programs; consideration of different settings (especially the availability of renewable energy sources) and technological solutions; exploring job creation and boosting local/regional economy
- Exploration of strategies in existing urban structures: combining revitalization and greening strategies; involving stakeholders with a special focus on landowner and citizen participation

Based on these preliminary observations, the next activities will focus on the following aspects:

- Integration of key findings into PED Program activities
- Developing the PED Booklet as strategic tool for the PED Program
- Updating the template to collect project descriptions
- Updating the PED Booklet with more cases across Europe

**Author Contributions:** Conceptualization: C.G. and S.B.; methodology: S.B. and C.G.; software: S.B., S.T., and C.G.; validation: S.B. and C.G.; formal analysis: S.B. and C.G.; investigation: S.B. and C.G.; resources, S.B. and C.G.; data curation, S.B. and S.T.; writing—original draft preparation, S.B.; writing—review and editing, C.G.; visualization, S.B.; supervision, C.G.; project administration, C.G.; funding acquisition, C.G. All authors have read and agreed to the published version of the manuscript.

**Funding:** This research received no external funding.

**Acknowledgments:** The authors would like to thank all project representatives who provided data on the projects presented in the PED Booklet, which served as basis for this analysis. Furthermore, the authors would also like to acknowledge the support of SET Plan Action 3.2 Italian National Delegate Paola Clerici and the work of the PED Program Management, specifically Margit Noll, Hans-Günther Schwarz, Robert Hinterberger, and Susanne Meyer, and the contributions of the PED Program Stakeholders Group, specifically EERA JPSC, regarding overall conceptualization of the PED Program and the elaboration of the PED Framework, which is a fundamental basis for the analysis of the data provided.

**Conflicts of Interest:** The authors declare no conflict of interest.

## References

1. Pichler, P.P.; Zwickel, T.; Chavez, A.; Kretschmer, T.; Seddon, J.; Weisz, H. Reducing Urban Greenhouse Gas Footprints. *Sci. Rep.* **2017**, *7*, 14659. [CrossRef] [PubMed]
2. Sustainable Development Goal 11: Make Cities and Human Settlements Inclusive, Safe, Resilient and Sustainable. Available online: https://sustainabledevelopment.un.org/sdg11 (accessed on 5 June 2019).
3. SET-Plan Action 3.2 on Smart Cities and Communities. Available online: https://setis.ec.europa.eu/system/files/setplan_smartcities_implementationplan.pdf (accessed on 12 October 2020).
4. JPI Urban Europe, Positive Energy Districts (PED). Available online: https://jpi-urbaneurope.eu/ped/ (accessed on 12 October 2020).
5. Framework Definition for Positive Energy Districts and Neighbourhoods. March 2020. Available online: https://jpi-urbaneurope.eu/app/uploads/2020/04/White-Paper-PED-Framework-Definition-2020323-final.pdf (accessed on 12 October 2020).
6. Alpagut, B.; Akyürek, Ö.; Mitre, E.M. Positive Energy Districts Methodology and Its Replication Potential. *Proceedings* **2019**, *20*, 8. [CrossRef]
7. Ahlers, D.; Driscoll, P.; Wibe, H.; Wyckmans, A. Co-Creation of Positive Energy Blocks, 1st Nordic conference on Zero Emission and Plus Energy Buildings. *Earth Environ. Sci.* **2019**, *352*, 012060.
8. Magrini, A.; Lentini, G.; Cuman, S.; Bodrato, A.; Marenco, L. From nearly zero energy buildings (NZEB) to positive energy buildings (PEB): The next challenge—The most recent European trends with some notes on the energy analysis of a forerunner PEB example. *Dev. Built Environ.* **2020**, *3*, 100019. [CrossRef]
9. Europe towards Positive Energy Districts, First Update. Available online: https://jpi-urbaneurope.eu/app/uploads/2020/03/PED-Booklet-Update-Feb2020_.pdf (accessed on 1 October 2020).
10. Berker, T.; Woods, R. Identifying and addressing reverse salients in infrastructural change. The case of a small zero emission campus in Southern Norway. *Int. J. Sustain. High. Educ.* **2020**, *21*. [CrossRef]
11. Cachat, E.T.; Grynning, S.; Thomsen, J.; Selkowitz, S. Responsive building envelope concepts in zero emission neighborhoods and smart cities—A roadmap to implementation. *Build. Environ.* **2019**, *149*, 446–457. [CrossRef]
12. Backe, S.; Kvellheim, A.K. *Zero Emission Neighborhoods—Drivers and Barriers towards Future Development*; ZEN Report 22; NTNU/SINTEF: ZEN, Norway, 2020.
13. Bertelsen, S.; Livik, K.; Myrstad, M. D2.1 Report on Enabling Regulatory Mechanism to Trial Innovation in Cities. +CityxChange, Work Package 2, Task 2.1; +Cityxchange, H2020, Grant Agreement No. 824260. Final delivery date: 31 July 2019.
14. Shnapp, S.; Paci, D.; Bertoldi, P. *Enabling Positive Energy Districts across Europe: Energy Efficiency Couples Renewable Energy*; JRC Technical Report; Publications Office of the European Union: Luxemburg, 2020.

15. Vimpari, J.; Seppo, J. Estimating the diffusion of rooftop PVs: A real estate economics perspective. *Energy* **2019**, *172*, 1087–1097. [CrossRef]

*Article*

# Testing Platforms as Drivers for Positive-Energy Living Laboratories

**Silvia Soutullo [1,*], Laura Aelenei [2], Per Sieverts Nielsen [3], Jose Antonio Ferrer [1] and Helder Gonçalves [2]**

1 Department of Energy, CIEMAT, 28040 Madrid, Spain; ja.ferrer@ciemat.es
2 Department of Renewable Energy and Energy Efficiency, LNEG, 1649-038 Lisbon, Portugal; laura.aelenei@lneg.pt (L.A.); helder.goncalves@lneg.pt (H.G.)
3 Department of Technology, Management and Economics, DTU, 2800 Lyngby, Denmark; pernn@dtu.dk
* Correspondence: silvia.soutullo@ciemat.es; Tel.: +34-913466305

Received: 30 September 2020; Accepted: 22 October 2020; Published: 27 October 2020

**Abstract:** The development of city-driven urban laboratories was considered a priority by the European Commission through Action 3.2 of the Strategic Energy Technology Plan. In this context, positive-energy districts laboratories could take the role of urban drivers toward innovation and sustainability in cities. These urban labs can provide real-life facilities with innovative co-creation processes and, at the same time, provide testing, experimenting, and prototyping of innovative technologies. In this scope, the authors of this work want to share the very first results of an empirical study using the testing facilities provided by the members of the Joint Program on Smart Cities of the European Energy Research Alliance as positive-energy districts laboratories. Six climatic regions are studied as boundary conditions, covering temperate and continental climates. Four scales of action are analyzed: Building, campus, urban, and virtual, with building and campus scales being the most frequent. Most of these laboratories focus on energy applications followed by networks, storage systems, and energy loads characterization. Many of these laboratories are regulated by ICT technologies but few of them consider social aspects, lighting, waste, and water systems. A SWOT analysis is performed to highlight the critical points of the testing facilities in order to replicate optimized configurations under other conditions. This statistical study provides guidelines on integration, localization, functionality, and technology modularity aspects. The use of these guidelines will ensure optimal replications, as well as identify possibilities and opportunities to share testing facilities of/between the positive-energy district laboratories.

**Keywords:** positive energy districts labs; testing facilities; factsheets; statistical studies

## 1. Introduction

Smart cities and communities have been identified as key challenges for achieving the energy efficiency targets for 2020 and 2050 according to the European Commission initiative formulated in the Strategic Energy Technology Plan Action 3.2 (SET Plan Action 3.2) [1]. The transition pathways need an integrated and inclusive approach that considers sustainability strategies, societal needs, and business opportunities on various levels.

Currently, 55% of the world's population lives in urban areas, a number that is expected to increase to 68% by 2050 [2]. At the same time, these urban areas are responsible for 71–76% of $CO_2$ emissions and 67–76% of global energy use [2]. These cities have been conceived as habitable, work, and leisure areas, but they also involve serious environmental, energy, social, and economic problems. New multi-criterion approaches are needed to evaluate and plan a sustainable energy transformation of the urban environments [3], as defined in the United Nations 2030 Agenda [4] and the Green Deal of the European Commission [5].

Seventy-five percent of the 210 million European buildings are not energy-efficient, and it is estimated that 75% to 85% of these buildings will be in use by 2050 [6]. The urban building stock is getting older and requires refurbishment measures that are energy-efficient, sustainable, and cost-effective. These measures should be adapted to the change in climate based on real weather data [7,8]. The combination of passive techniques to reduce the building energy demands [9,10] with techniques to minimize the primary energy consumption [11,12] represents a path toward decarbonization and improvement in energy and environmental impacts in cities. Thermal comfort, air quality, and appropriate light and acoustic levels should be considered inside buildings [13–15]. However, the building energy optimization should be accompanied by energy savings in cities. This concept includes aspects such as energy storage, flexibility, or polygeneration [16–18]. Urban approaches need to be carried out by means of optimized planning that considers energy and environmental aspects [19,20]. Sustainable urban mobility is an imperative need as a source of congestion [21]. The measures proposed differ from city to city depending on the urban planning, citizens, and stakeholders [22]. The use of Information and Communication Technologies (ICT) makes it possible to evaluate the information coming from urban transport and identifies different aspects that allow the improvement in its efficiency and sustainability [23]. Furthermore, other challenges such as the influence of air quality [24] and its measurement [25], urban livability [26], system integration of all urban flows [27], and lack of citizen involvement [28,29] are the actual real challenges toward sustainable urban transformation and energy transition.

Over many decades, the European Commission has been dedicating attention on research challenges for development and demonstration, isolated measures, and technologies rather than proposing an integrated approach where technologies and processes would work together in the urban environment. This situation changed in 2004 when the European Commission started strongly supporting this integration approach at the city level, specifically with the Framework Programme Research and Innovation 6 funding of the CONCERTO projects [30]. In line with this challenge, the Horizon 2020 framework has provided smart cities and communities lighthouse projects with approximately €500 million in funding for 2014–2020 [31].

In 2015, the SET-Plan Action 3.2 Smart Cities and Communities [32] promoted by the European Commission identified specific targets related to buildings. These include the development of the necessary interfaces to connect net-zero-energy buildings [33,34] and positive-energy buildings [35] to zero-energy blocks and districts [36,37]. This new concept integrates renewable energy systems, ICT solutions, smart meters, smart appliances, smart energy management, empowering tenants, improving planning management, and developing tools for scalable integrated design. In 2010, the Joint Programming Initiative (JPI) Urban Europe [38] was created to develop knowledge, tools, and platforms to help European cities on their way to sustainable urban transitions. This initiative connects public authorities, civil society, scientists, innovators, companies, and industry with the aim of providing a new environment for research and innovation in urban issues.

The European Energy Research Alliance (EERA) is the largest energy research community in Europe [39] made up of 250 universities and public centers from 30 European countries. The EERA mission is to catalyze energy research to achieve a climate-neutral society by 2050. With this aim, a wide range of low-carbon technologies that enable a clean energy transition are analyzed. To work on the priority lines of low-carbon technologies defined by the Strategic Energy Technology Plan, the EERA has created 17 Joint Programs (JP). One of these JP is focused on Smart Cities [40], contributing to research and innovation of Positive-Energy Districts (PED). Within this framework, the EERA Joint Programme on Smart Cities (EERA JP SC) has implemented the work established in the Temporary Working Group of the SET-Plan Action 3.2 on PEDs [32]. One of the strategic solutions and relevant actions proposed to respond to these challenges and targets is the use and development of a network of infrastructures for industrial research and demonstration through module 2. This strategic module assesses the development of city-driven Positive-Energy Districts Laboratories (PED Labs). One of the objectives of this module is to identify research testing platforms across countries in Europe that can be used for developing and testing integrated solutions and should work as drivers for local communities, districts, and cities.

This article makes a first mapping of the existing urban living laboratories in Europe in order to identify the resources available to deal with cross-cutting issues. To this end, only testing installations of the members of the EERA JP SC have been evaluated. General and more specific statistical studies have been developed to identify the main characteristics of these facilities. Finally, a SWOT analysis has been carried out to highlight the strengths, weaknesses, opportunities, and threats reached by the studied sample.

## 2. Materials and Methods

The implementation and monitoring of new technological and innovative solutions, governance models, socio-economic models, citizen participation or environmental issues highlight the most influential fluxes for transforming our neighborhoods. Based on this, PED Labs should consider many aspects to optimize different urban configurations under several conditions. These laboratories have to:

- Be flexible to allow the integration of different solutions.
- Take into account energy, social, environmental or economical aspects.
- Monitor the performance of a district solution under real conditions of use.
- Quantify the main fluxes that contribute to the studied configuration.

To identify the resources available in Europe that allows for evaluating integrative urban configurations, a first mapping of the existing urban living laboratories is needed.

This article makes a first mapping of the urban testing facilities that have been provided by the members of EERA JP SC. The methodology used has been divided into several stages. The first stage is to evaluate the capabilities of the existing urban infrastructures. With this aim, a factsheet has been developed to quantify the main characteristics of the living laboratories submitted. This factsheet describes the location of the installation, the field of action, the testing facilities available, the plant size, the main capacities of the installation, and the person of contact.

Figure 1 represents two examples of the factsheets describing the testing infrastructures provided by the EERA JP SC members. On the right side, the testing facility of National Laboratory of Energy and Geology (LNEG) in Lisbon (Portugal) is shown, while the left side corresponds to the testing laboratory of the Centre for the Development of Renewable Energy (CEDER-CIEMAT) in Soria (Spain).

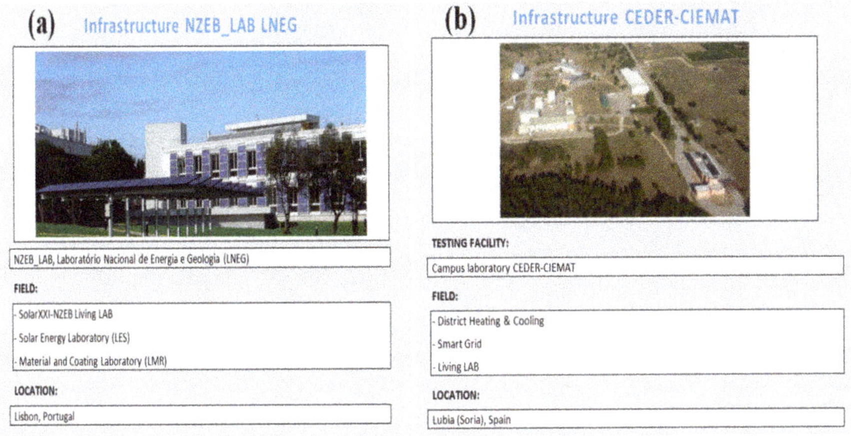


**Figure 1.** Examples of factsheet developed by module 2 of the European Energy Research Alliance Joint Programme on Smart Cities (EERA JP SC). (**a**) NZEB lab in the facilities of LNEG, (**b**) campus lab in the facilities of CEDER-CIEMAT.

The second stage consists of a statistical study of the laboratories described in the submitted factsheets. This survey allows the identification of the frequency of boundary conditions, scale of action, facility application, and technologies available in these laboratories. These data have been disaggregated by each scale of action obtained in the studied sample. Finally, a SWOT analysis has been carried out to identify the strengths, weaknesses, opportunities, and threats found in the studied laboratories. This evaluation highlights the resources available, the critical points, the benefits produced, and the barriers reached in the operation of these urban living-labs.

## 3. Results

Once the information of the infrastructures submitted by the members of the EERA JP SC has been compiled, statistical studies have been performed to identify the main characteristics of these facilities. These evaluations provide information regarding climate conditions, the laboratory scale of action, facilities application, equipment, and technologies or experimental devices. This information can be used for development, testing, and optimizing different technologies and solutions, detecting what is available, and identifying what would be needed to create or replicate a new PED lab.

### 3.1. Case Studies

The first approach consists of sixteen testing platforms and existing facilities provided by the EERA JP SC members. These cases represent six European countries: Portugal, Spain, Belgium, Czech Republic, Finland, and Norway. Despite being a small sample, different regions are covered: North, south, and central Europe, as it can be seen in Figure 2. More cases should be included in these statistics in order to consider different climate conditions.

**Figure 2.** European map highlighting the initial testing facilities studied by module 2 of EERA JP SC [Source: Google Maps].

Sixteen facilities have been compiled from Norwegian University of Science and Technology (NTNU); Spanish Centre for Energy, Environment and Technology Research (CIEMAT); Spanish Centre for the Development of Renewable Energy (CEDER-CIEMAT); Public University of Navarre & Campus

Iberus (UPNA-Campus Iberus); University of Zaragoza & Campus Iberus (UniZar-Campus Iberus); eXiT research group from the Univeristy of Girona (ExiT RG-UdG); National Laboratory of Energy and Geology from Portugal (LNEG), Nieuwe Dokken district from Belgium; Energyville Centre from Belgium, Catalonian Institute for Energy Research (IREC), University Centre for Energy Efficient Buildings from the Technical University of Prague (UCEEB) and Technical Research Center of Finland Ltd (VTT). The study of these factsheets can help to highlight the capabilities available by the EERA JP SC members [33] to assess different aspects of PEDs across Europe.

The main characteristics of these infrastructures are summarized in Table 1 and are based on the results provided by the factsheets submitted.

**Table 1.** Name, center, location, country, scale of action, and contact person defined for the sixteen cases submitted in the first approach.

| Infrastructure | Center | Location | Country | Scale of Action | Corresponding Person |
|---|---|---|---|---|---|
| ADRESSAPARKEN | NTNU | Trondheim city center | Norway | Park, community | andrew.perkis@ntnu.no |
| CEDER | CEDER-CIEMAT | Lubia, Soria | Spain | Research campus | raquel.ramos@ciemat.es |
| ED70 | CIEMAT | Madrid | Spain | Building | Ja.ferrer@ciemat.es |
| UPNA | Campus Iberus | Pamplona, Navarra | Spain | University campus | jmolina@unizar.es |
| UniZar | Campus Iberus | Zaragoza, Aragón | Spain | University campus | jmolina@unizar.es |
| UdG. | eXiT RG. | University of Girona, Cataluña | Spain | University campus | roberto.petite@udg.edu/ robert.rusek@udg.edu |
| NZEB_LAB | LNEG | Lisbon | Portugal | Building | laura.aelenei@lneg.pt |
| NIEUWE DOKKEN | Nieuwe Dokken | Ghent | Belgium | District | denieuwedokken.be/contact |
| THOR PARK | Energyville | Genk | Belgium | Community/park | daan.six@energyville.be |
| SEILAB ENERGY | IREC | Barcelona, Cataluña | Spain | Test building | Jsalom@irec.cat |
| SMARTLAB | IREC | Tarragona, Cataluña | Spain | facility | ccorchero@irec.cat |
| UCEEB | Technical University Prague | Buštěhrad | Czech Republic | University test building/facility | michal.kuzmic@cvut.cz |
| ZEB LIVING LAB- | NTNU | Campus Gløshaugen Trondheim | Norway | Building | Kristian.skeie@ntnu.no |
| SPARCS Project | VTT | Leppävaara District, Espoo | Finland | Virtual Power Plant | jaano.juhmen@siemens.com |
| SPARCS Project | VTT | Lippulaiva (Espoonlahti), Espoo | Finland | District | kaisa.kontu@citycon.com |

### 3.2. Climate Conditions

The Köppen–Geiger classification [41] is used to highlight the climate representativeness of the sixteen studied cases. Figure 3 shows the Köppen–Geiger classification registered in Europe [42].

The study sample covers temperate and continental climate conditions and excludes polar and tropical conditions. Six climate regions have been identified:

- Zone BSk is defined as cold and semi-arid climate. These zones are characterized by warm to hot dry summers and cold winters. The precipitation patterns for regions with higher latitudes correspond to dry winters and wetter summers. The precipitation patterns for regions with lower latitudes correspond to dry summers, relatively wet winters, and wetter springs and autumns. Only the city of Zaragoza in Spain is available in this climatic zone.
- Zone Csa is defined as hot-summer Mediterranean climate. These zones are characterized by hot and dry summers and mild and wet winters. Four Spanish regions are available in this climatic zone: Barcelona, Girona, Madrid, and Tarragona.

- Zone Cfb is defined as temperate oceanic climate. These zones are characterized by moderate temperature along the year, absence of a dry season and constant precipitation especially during the colder months. Three countries are available in this climatic zone: Spain (Soria and Pamplona), Belgium (Dilsen-Stokkem, Ghent, and Genk), and the Czech Republic (Buštěhrad).
- Zone Csb is defined as warm-summer Mediterranean climate. These zones are characterized by warm and dry summers and wet winters with temperatures that vary from mild to chilly climates. Only the city of Lisbon in Portugal is available in this climatic zone.
- Zone Dfb is defined as warm summer and humid continental climate. These zones, known as hemiboreal zones, are characterized by warm to hot and humid summers and cold winters (sometimes severe cold in northern areas). Precipitation is distributed throughout the year. Only the city of Espoo in Finland is available in this climatic zone.
- Zone Dfc is defined as subarctic climate. These zones are characterized by long and very cold winters and short summers with temperatures that vary from cold to mild climates. In these regions, there is no dry season and there are no significant seasonal differences in precipitation. Only the city of Trondheim in Norway is available in this climatic zone.

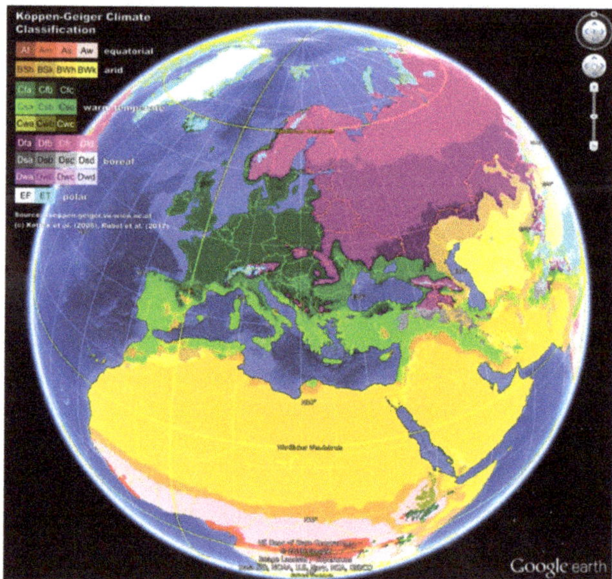


**Figure 3.** European map of Köppen–Geiger classification developed by the Climate Change and Infectious Diseases Group [42].

The percentage distribution of the climate zones registered by the case studies is shown in Figure 4. With this first sample, the most frequent conditions are characterized by temperate climate conditions (Cfb zone), followed by hot summer and mild winter areas (Csa zone). On the opposite side, cold and semi-arid areas (BSk zone) and warm summer with mild winter climates (Csb zone) are the most infrequent zones.

The variability in the available climates allows the analysis of numerous urban configurations, taking into account all kinds of renewable and low-carbon technologies. Thermal conditioning strategies for both exterior spaces and interior buildings cover a wide range of measures, from the maximization of solar gains or optimal distribution for the wind flows to adiabatic cooling or optimized installation of shading elements. Unfortunately, this sample does not cover regions with more extreme climates such polar or tropical conditions.

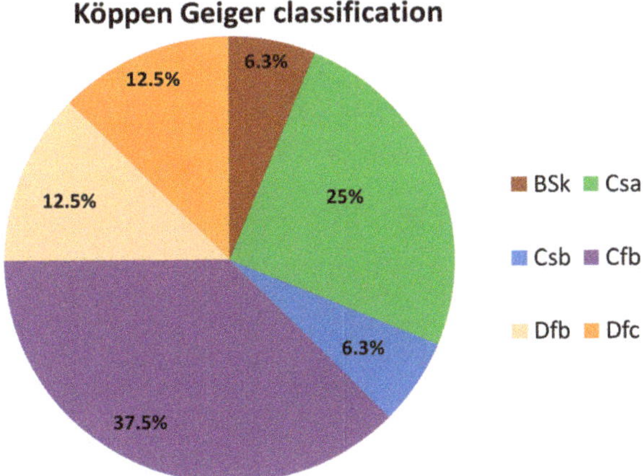


**Figure 4.** Distribution of the Köppen–Geiger climate classification obtained for the studied cases.

### 3.3. *Scale of Action*

According to information provided by the factsheets, four scales of action have been identified for the studied cases: Building, campus, urban, and virtual. The differences between them are based on the dimensions, boundary conditions, and the energy fluxes that can be evaluated by these facilities. The first three scales of action have geographical limits for their facilities, while the fourth does not have these types of restrictions. Figure 5 shows the percentage distribution of the scales of action obtained for the studied cases. As can be seen, the campus and the building scale are the most common, both reaching more than 62% of the studied cases. These installations allow the evaluation and controlling of technical solutions and methodologies in two scales that, once optimized, can be implemented into the urban infrastructures. On the other side, virtual laboratories are the most infrequent cases with a percentage that does not exceed 13%. These installations allow the evaluation of energy exchanges that are not subject to geographic limits.

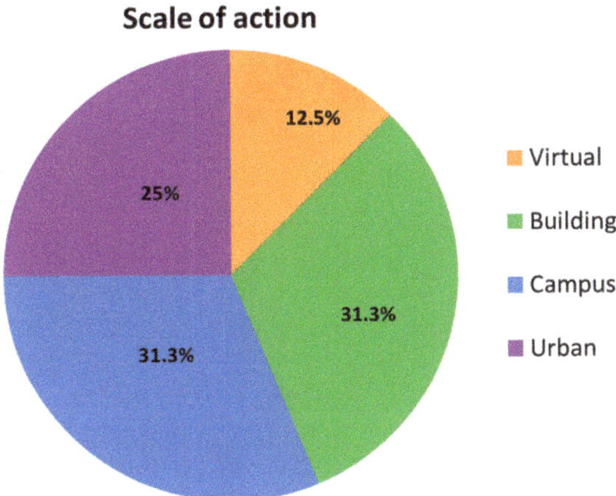


**Figure 5.** Distribution of the action scale defined for the laboratories studied.

The scales of action defined for the laboratories determine the type of experiments than can be done. Different control levels are identified regulating the fluxes that contribute to the urban balance. The principal characteristics of these scales of action are:

- Building scale (green area). These facilities consider the building as a very small-scale urban experimental laboratory where energy exchanges between the building and its surrounding environment can be controlled. In these installations, the boundary conditions, as well as the building energy fluxes, are evaluated to identify the relevant flows that will affect the district scale. The ZEB lab is shown in Figure 6 as an example of the building scale.

**Figure 6.** Building ZEB Lab provided by NTNU (Norway).

- Campus scale (blue area). This scale represents a testing district facility that operates under controlled conditions. These laboratories consider some elements of the district model such as buildings and conditioning networks, in which the energy fluxes are evaluated in a controlled manner. Several batteries of experiments can be performed to reproduce specific situations. The facilities of EnergyVille placed in Belgium are shown in Figure 7 as an example of this scale.

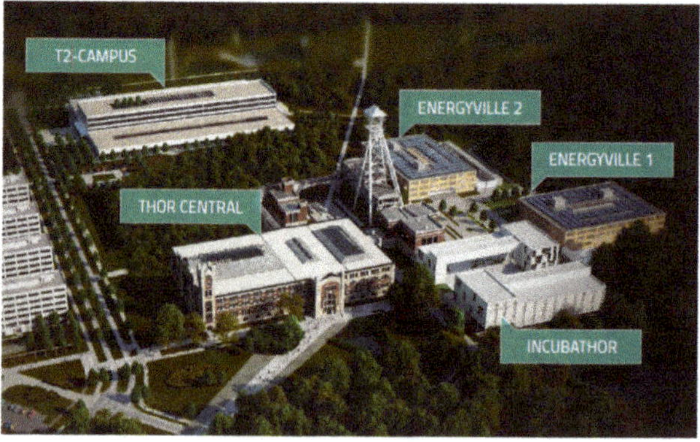


**Figure 7.** Diagram for the campus Thor Lab provided by EnergyVille (Belgium).

- Urban scale (purple area). This scale represents urban testing facilities operating under real conditions of use. Buildings, thermal networks, electrical networks or open spaces are elements of these laboratories. In these laboratories, all the urban interactions between people, buildings, and systems are represented, but not all of them can be controlled and calculated. The Lippulaiva lab is shown in Figure 8 as an example of the urban scale.

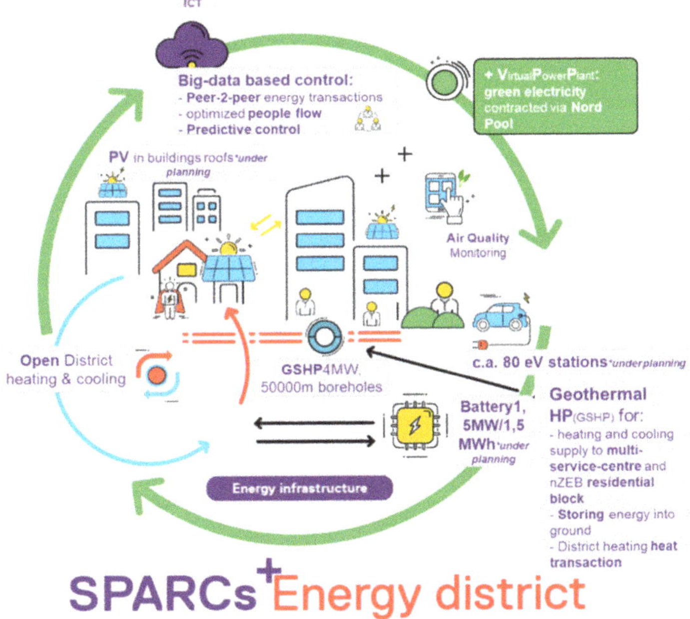


**Figure 8.** Diagram for the urban Lippulaiva Lab provided by VTT (Finland).

- Virtual scale (orange area). This scale represents a laboratory that can operate under real or simulated conditions of use. These facilities allow the evaluation of several urban fluxes, but high computing requirements are needed. Real measurements of different elements must feed these platforms to represent the urban energy fluxes correctly. The semi virtual SEILAB facility is shown in Figure 9 as an example of this scale.

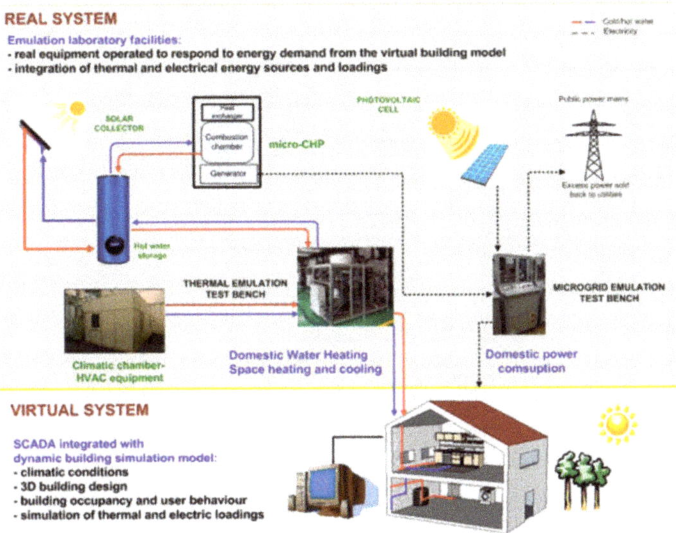


**Figure 9.** Diagram for the virtual SEILAB facility provided by IREC (Spain).

### *3.4. Facility Application*

Based on the information provided by the factsheets, several applications for the testing facilities are identified. These uses define the type of experiments, the capacity of the laboratories, and the urban flows that can be assessed. These laboratories analyze different aspects of the urban balance, such as: Energy factors, environmental characteristics, information and communication technologies, or social aspects. The main applications identified are:

- Social aspects. The interactions between humans and their environment can be analyzed as an urban issue in the existing facilities.
- ICT/control. These laboratories assess the performance of control systems and technologies of information and communication.
- Outdoor climate conditions. The ambient climate conditions of the district can be evaluated through experimental devices.
- Indoor climate conditions. The ambient conditions inside the buildings can be monitored through the integration of experimental devices.
- Energy loads. The building performance can be quantified through the energy loads calculation. Two types can be evaluated: Thermal and electrical loads.
- Electrical vehicle. The interactions between electrical vehicles, buildings, and grids can be assessed through electrical mobility facilities.
- Lighting systems. The operation of artificial lighting systems produced in the district can be analyzed and regulated through different equipment.
- Energy networks. The energy interaction between the generation sources, distribution networks, and consumption points can be evaluated through the use of different devices. Two district types can be evaluated: Thermal (heating and cooling) and electrical networks.
- Storage elements. The fluctuations produced between the energy production and the demand side can be evaluated through the energy performance of the storage systems. Two types can be evaluated: Thermal and electrical storage.
- Water systems. The performance of water recovery systems or treatment of water can be assessed by means of different devices in the existing facilities.
- Waste treatments. These laboratories evaluate different solutions to minimize the urban waste and increase the environmental conditions.

The general statistics of the facility applications obtained for all the studied cases are shown in Figure 10. This figure represents a histogram with the frequencies of occurrence obtained for each application. The majority of the studied installations evaluate energy loads, energy networks, especially the electrical grids, and energy storage systems. Most of these laboratories have control or ICT systems to evaluate and optimize their operation. Many installations analyze the electrical vehicle and monitor ambient conditions, especially outdoors situations. On the opposite side, few installations assess lighting devices, social aspects, waste solutions, and water systems.

### *3.5. Characteristics of Laboratories according to the Scale of Action*

Once the possible applications of the testing infrastructures have been defined in a general way, the most common ones at each scale of action are identified, specifying the technologies that are involved in the experimental processes.

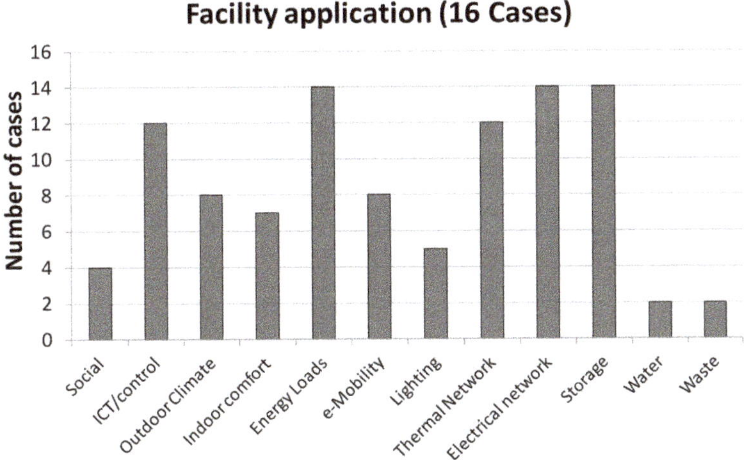


**Figure 10.** Histogram of occurrence obtained for the facility applications of the studied cases.

### 3.5.1. Building Scale

The smallest scale of action is the building scale, in which the exchanges between the building and the urban environment are regulated in a controlled manner. There are five testing facilities in the studied sample, located in four climate zones: One placed in the Cfb zone, one placed in the Csb zone, one placed in the Dfc zone, and two placed in the Csa zone. Figure 11 shows the technologies installed in the studied cases at the building scale. All of the installations evaluated at this scale of action assess the building energy loads, regulate the operation with ICT/control devices, and have electrical systems and thermal systems. The majority evaluate outdoor and indoor conditions and have storage devices. Three installations have implemented ventilation techniques and artificial lighting into the building. Finally, only one laboratory has water treatment and electrical vehicle assessment. There are other technologies installed inside these building laboratories such as phase-change materials or ventilated façades.

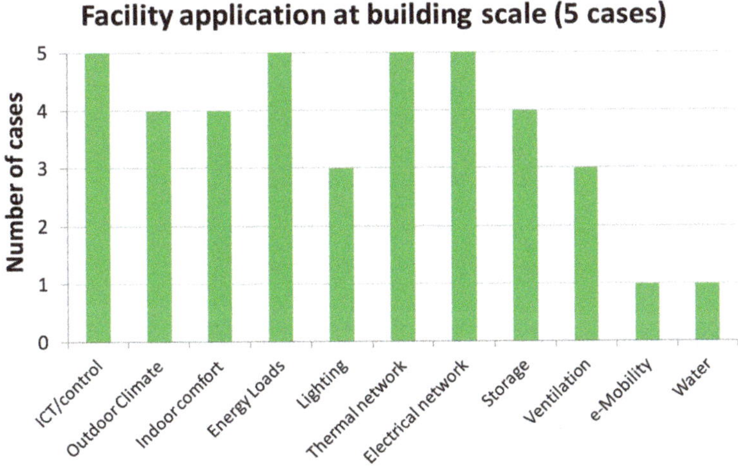


**Figure 11.** Histogram of occurrence obtained for the facility applications at the building scale.

Attending to the technologies installed inside these controlled buildings, solar thermal collectors, photovoltaic panels, biomass boilers, wind turbines, heat exchangers, ground exchangers, energy storage systems, absorption chillers or water systems can be used to develop different batteries of experiments. Figure 12 shows the technologies installed in the studied cases. The installation of solar photovoltaic panels is the most common technology reached in the buildings studied. It is followed by solar thermal collectors. Heat exchangers, storage elements or cooling systems have been obtained with great percentages. No wind turbines or biomass boilers have been found in the test-buildings studied. In addition, other technologies are available in the building sample such as cogeneration systems or the organic Rankine cycle.

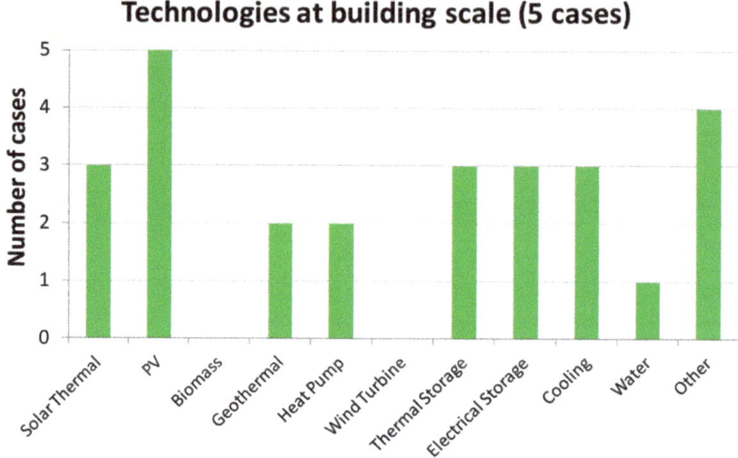


**Figure 12.** Occurrence of the various technologies installed at the building scale.

### 3.5.2. Campus Scale

An intermediate scale of action is the campus scale, in which the urban exchanges are regulated in a controlled manner. There are five testing facilities in the studied sample, located in three climate zones: Three placed in the Cfb zone, one placed in the Csa zone, and one placed in the BSk zone. Figure 13 shows the histogram of occurrence obtained for the facility applications found at the campus scale. All these laboratories have control or ICT systems to optimize the use of the installations as well as energy storage systems. The majority of them have energy networks, especially electrical. Most of these laboratories integrate buildings into the campus installation and evaluate the ambient conditions, especially indoors. On the other side, only two laboratories analyze the electrical mobility and monitor outdoor conditions. There is no installation that considers water systems.

Once the facility applications have been identified for the studied cases, the technologies installed should be highlighted, shown in Figure 14. This graph represents the histogram of occurrence reached by installations available at the campus scale. The majority of these testing laboratories have thermal and electrical storage systems.

Most of the campus laboratories have installed solar photovoltaic panels, as well as other technologies such as an organic Rankine cycle, Stirling engines, fuel cells or LED lighting. Two laboratories have integrated geothermal and cooling systems. Solar thermal collectors, biomass boilers, heat pumps or wind turbines have been obtained in one case.

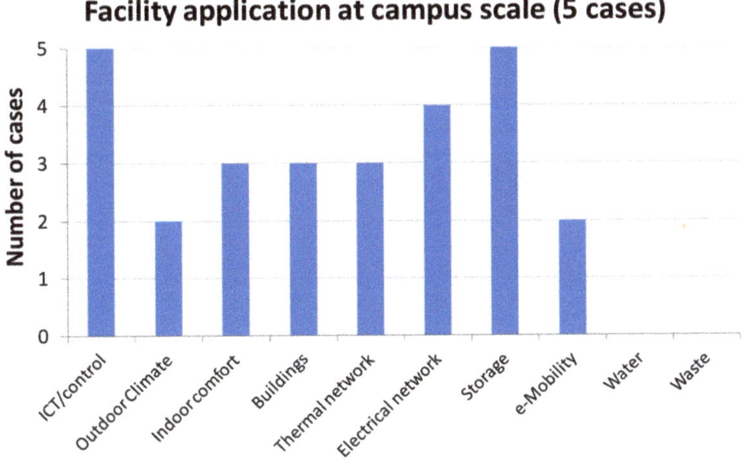


**Figure 13.** Histogram of occurrence obtained for the facility applications at the campus scale.

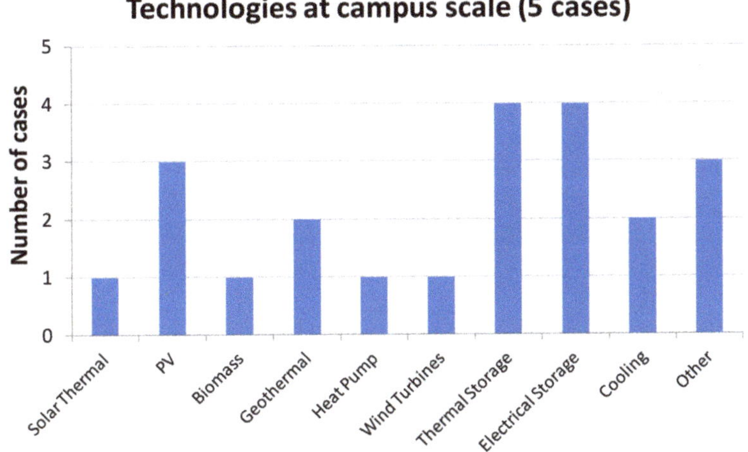


**Figure 14.** Histogram of occurrence obtained for the technologies installed at the campus scale.

### 3.5.3. Urban Scale

The urban scale of action evaluates the urban interactions between buildings, technologies, and people, but not all of them are analyzed and regulated. There are four testing facilities in the studied sample, located in three climate zones: Two placed in the Cfb zone, one placed in Dfc zone, and one placed in the Dfb zone. By analyzing the factsheets provided by the EERA JP SC members, the histogram of occurrence for the facility applications is obtained and plotted in Figure 15. All the urban laboratories studied have considered social areas, and most of them have included buildings, thermal and electrical grids, electrical mobility, and the participation of citizens. Half of these facilities have green areas, ICT or control systems or have considered the treatment of the urban waste. Very few laboratories have included lighting systems, water treatments, and monitoring of urban outdoor conditions.

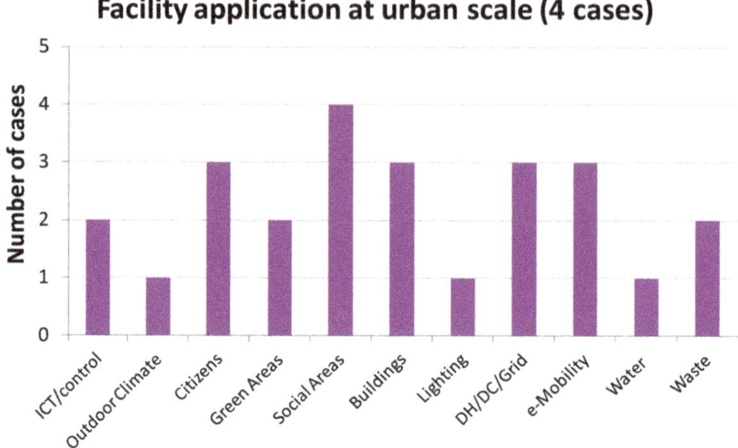


**Figure 15.** Histogram of occurrence obtained for the facility applications at the urban scale.

The technologies included in these urban laboratories are photovoltaic solar panels, biogas boilers, geothermal heat pumps, thermal and electrical storage systems, and waste treatment. The percentage of occurrence of these technologies is shown in Figure 16. The most implemented technologies in the studied urban laboratories are photovoltaic panels and electrical storage. Half of these installations have included waste treatments. Only in one case, a biogas boiler, a geothermal heat pump or thermal storage systems have been installed. However, some of these laboratories have included efficient buildings, cogeneration plants or open-source technical infrastructures.

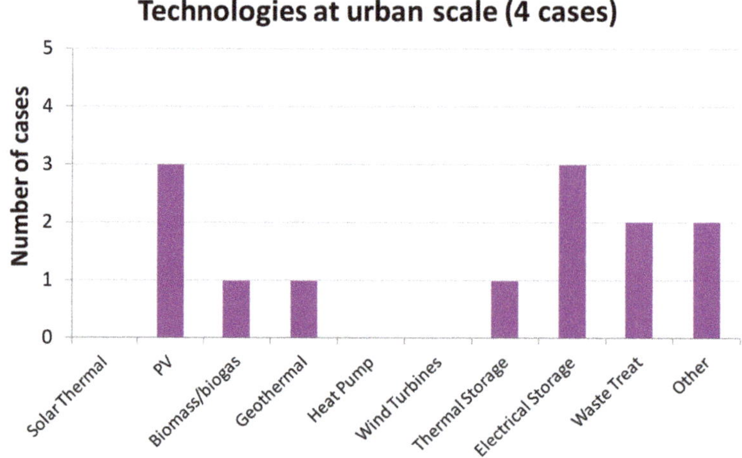


**Figure 16.** Histogram of occurrence obtained for the technologies installed at the urban scale.

3.5.4. Virtual Scale

The virtual scale of action evaluates the urban interactions in a simulated environment based on real data. There are two testing facilities in the studied sample, located in the Csa and Dfb climate zones. Both testing laboratories evaluate outdoor conditions, building energy loads, and electrical vehicles. These labs assess the performance of several thermal and electrical technologies, as well as the operation of storage systems. All these energy flows are modeled and controlled through the use of ICT systems. Only one laboratory analyzes ambient conditions inside the buildings and only one laboratory considers lighting systems. Any virtual laboratory has integrated water treatment systems. All the applications found by these two facilities are shown in Figure 17.

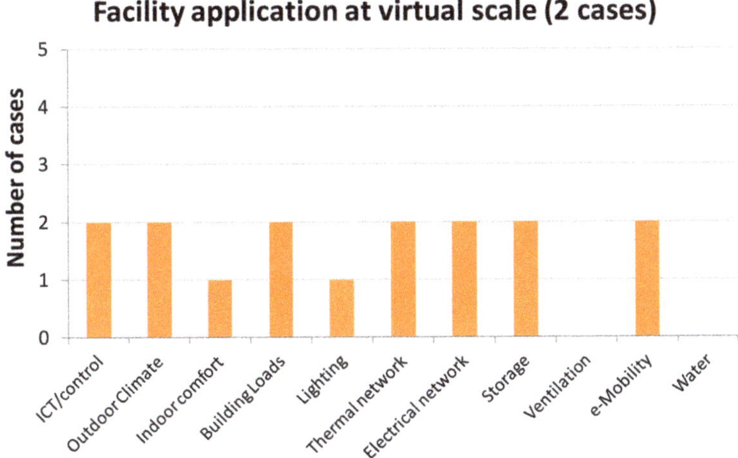


**Figure 17.** Histogram of occurrence obtained for the facility applications at the virtual scale.

The most common technologies evaluated in these laboratories are: Photovoltaic solar panels, electrical storage, and virtual power plants, as can be seen in Figure 18. On the other hand, no solar thermal plants, ground exchanges, wind turbines or absorption chillers are available in these testing facilities. The laboratory placed in the Csa zone has electrical and heat test benches to emulate heat and electrical sources. It also has heat pumps, storage tanks, and gas boilers. The laboratory placed in the Dfb zone has district heating supplied by the renewable heat from a bio oil plant, a decentralized energy production, and the inclusion of an e-mobility hub. All these flows are managed through a virtual power plant.

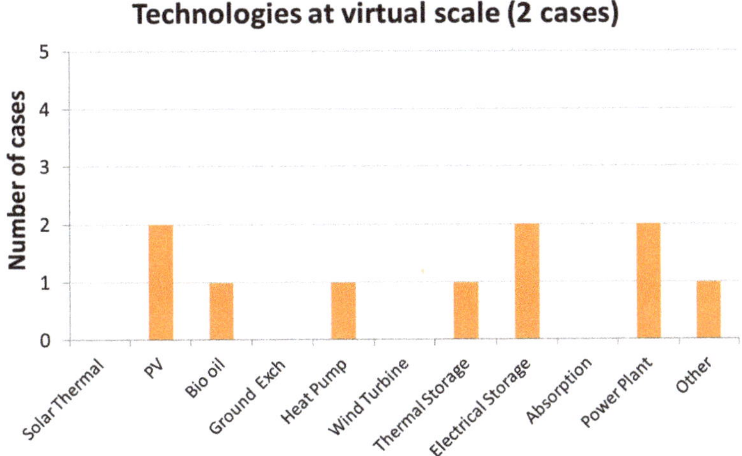


**Figure 18.** Histogram of occurrence obtained for the technologies installed at the virtual scale.

### 3.6. SWOT Analysis

In order to know the real situation of the available urban laboratories and plan a future strategy for PED Labs, a SWOT analysis is carried out. This study highlights the strengths, weaknesses, opportunities, and threats reached by the whole studied sample. This SWOT analysis highlights the main technical, administrative, economic, social, environmental or other aspects that may constitute a barrier (weaknesses, threats) or those that may promote (strengths, opportunities) the development and replication of urban laboratories. The identification of these points makes it possible to know the available resources and requirements needed when analyzing different PED configurations.

The analysis performed includes the benefits and barriers obtained for the complete sample of the sixteen laboratories, but a more specific SWOT analysis can be done for each scale of action.

From the sample of sixteen facilities provided by the EERA JP SC members located throughout Europe, the following strengths have been identified:

Testing platforms STRENGTHS

- High climatic representativeness of Europe.
- Availability of different scales of action for the studied laboratories, allowing the assessment of several urban flows.
- Availability of measurements in real conditions of use and in virtual conditions.
- Availability of different technologies for the use of laboratories: Energy, social, e-mobility, ...
- Possibility of evaluating urban configurations that integrate various types of uses.
- Availability of different types of buildings, energy systems, and storage systems.
- High representativeness of renewable technologies.
- High generation potential of renewable technologies.
- Availability of other types of technologies such as cogeneration or Stirling engines.
- Diversification of energy generation and consumption due to the competitiveness of generation technologies and energy efficiency measures.

High availability and versatility of control systems.

Based on the information provided by this first dataset of the existing facilities, the following weaknesses have been identified:

Testing platforms WEAKNESSES

- Few laboratories available for the analysis sample. To raise the representativeness of the study, more cases should be added.
- Absence of extreme weather conditions: Tropical (climate zone A) and polar (climate zone E).
- Necessity to expand the climate applicability of this study. More dry (B zones) and continental (D zones) regions should be included.
- More virtual laboratories are needed.
- Low availability of laboratories that include social aspects, water, and waste flows.
- More cases of some renewable technologies should be included into the sample: Solar thermal technologies, wind turbines or biomass boilers.
- Gap between theoretical models and real situations of certain systems due to the lack of validation based on real data.
- Technical aspects in order to optimize the management of a combined configuration.
- Administrative, regulatory, and normative barriers can be found for the optimal operation of the laboratory.
- Absence of experimental analysis protocols for high scales of action.

The following opportunities have been identified:

Testing platforms OPPORTUNITIES

- Quantification of several urban flows thanks to the availability of different scales of action laboratories.
- Evaluation of the energy balance for combined urban solutions.
- Integrated management optimization for the resources and services of a district.
- Development of virtual PED labs to assess configurations that integrate different urban aspects.
- Helping tools to increase the decarbonization of the districts through sustainable models with a favorable impact on the quality of life of citizens.
- Optimization of solutions that promote the efficient renovation of the building stock and the optimization of public infrastructures.
- Improvement in the comfort perception and the environmental conditions in urban districts.
- Evaluate, under controlled conditions, the joint participation of different stakeholders for the development of positive-energy districts.
- Creation of qualified jobs.

Finally, the following threats have been identified:

Testing platforms THREATS

- Deviation in the obtained results due to the absence of measure protocols for PED labs.
- More regulatory barriers for urban large-scale laboratories.
- Need for flexible and efficient networks connected to the living labs.
- Administrative problems for urban large-scale laboratories due to data protection laws.
- Necessary investment for the implementation of some technologies and infrastructures.
- Legal barriers to construct an urban large-scale laboratory.
- Difficulty of defining a large-scale laboratory due to the existing urban layout in cities.
- Social lack of knowledge for the benefits produced by urban large-scale laboratories.

## 4. Discussion

To achieve the necessary energy transition in cities, it is essential to increase the renewable systems integration into urban energy networks, and to push energy efficiency performance levels of the built environment significantly beyond the levels of current European building codes. The development of efficient solutions that can be replicated or scaled up to the city level requires the use of urban laboratories. However, given the complexity of the urban context, it is necessary to have several scales of action for these laboratories in order to quantify and control different flows that contribute to the urban balance. Building, campus, urban, and virtual scales are identified to characterize the urban configurations. The first three scales have geographical restrictions; however, the fourth scale can operate under real or virtual conditions of use.

The use of PED laboratories as an analysis tool allows the quantification of flows of different configurations and identification of the critical points and the conditions that take place. Several technologies can be integrated into the monitored urban configurations, maximizing the use of efficient systems and renewable technologies. To achieve homogeneous conclusions from these living-labs, experimental analysis protocols and testing studies should be proposed.

Under the framework of the EERA Joint Program on Smart Cities, a factsheet has been designed to compile all the information regarding the capabilities of the living laboratories from the members. A statistical study has been performed to identify the boundary conditions, the scale of action, the facility applications, and the technologies available in these laboratories. Despite having only sixteen laboratories, a good representation of climatic zones in Europe has been obtained. Based on the Köppen–Geiger climatic classification, the most frequent zones for the studied laboratories are Cfb and Csa, while the most infrequent zones are BSk and Csb. To expand the applicability to extreme conditions, installations placed in tropical and polar climate zones should be added. More than 62% of the studied laboratories are built under building and campus scales of action (both in equal percentages), 25% are built under urban scales of action, and only 12.5% are built as virtual PED labs. It is necessary to develop more virtual laboratories that allow the analysis of integrated urban situations. From this study sample, most of the living-labs are focused on energy applications with a high contribution of networks and storage systems, as well as energy loads characterization. Many laboratories have high versatility to operate the installations using control and ICT technologies. There are a high percentage of laboratories that manage e-mobility and ambient conditions. Nevertheless, few integrate lighting devices, social aspects, and waste and water systems into the global installations. More laboratories with nonenergy uses must be added to increase the applicability of the urban results.

Attending to each scale of action, a high diversification is observed in the generation, storage, distribution, and consumption of energy. Different types of technologies are available (renewable and nonrenewable) that allow the evaluation of several urban configurations. There is a high contribution of renewable technologies, with solar photovoltaic generation being remarkable. However, to evaluate the high potential of other renewable technologies in the urban context, more laboratories with solar thermal collectors, wind turbines or biomass boilers should be added. Finally, different types and uses of buildings are available to quantify the urban loads. This fact enables flexibility when modelling the urban consumption profile.

A SWOT analysis has been carried out to highlight the technical, economic, environmental, administrative or social aspects found in the total sample studied, but it must be taken into account that there are different types of applications according to the laboratory scale of action. This scale defines the limits of benefits and barriers obtained in the testing facilities.

Once the first mapping of the existing living labs has been carried out, it is necessary to go deeper and identify other cross-cutting issues such as existing incentives, financial aspects, local and national policies, business models, and city planning. The complexity of these interrelations requires new approaches to address the problems in a holistic and integrated manner, using virtual PED labs to do so. These virtual laboratories will be fed with data from different sources (Building Information Modeling files, smart meters, Geographic information system, satellites, sensors), allowing the application of smart solutions that combine several urban configurations and open new paths to assess the transition to clean energy systems in cities.

## 5. Conclusions

The development of positive energy districts should be based on an integrative approach that considers not only energy efficiency and renewable integration across the built environment, but also other integrative processes. Within the urban context, different sectors, systems, strategies, goals, rights, and interests have to be considered. Therefore, urban models should take into account energy flexibility, the interaction between buildings and energy systems, the integration of users and local communities, economic issues, social aspects, environmental conditions, and legal considerations. In addition, all urban configurations should be designed based on climate change mitigation, decarbonization, and the energy transition targets.

The optimal planning of PEDs requires the knowledge of available natural resources, existing technologies, and their performance under specific boundary conditions. For this reason, it is necessary to have urban laboratories that allow testing, under real conditions of use, the most effective and sustainable solution for each geographical area. The use of PED labs makes it possible to optimize urban solutions, minimizing the existing risks in the implementation of measures. To quantify the urban flows as a function of the boundary conditions, it is necessary to have laboratories that operate under different scales of action. These laboratories should be flexible to test different configurations, with a representative sample of different conditions and applications being necessary. The complexity of the urban interrelations requires new approaches to address a solution holistically, meet the targets of sustainability strategies on various levels, and contribute to urban transition pathways at large. To homogenize the results obtained by these living-labs, it is necessary to have experimental analysis protocols. These protocols must have key and quantifiable indicators, which serve as a guide-control throughout the monitoring process. This methodology will allow the development of virtual PED labs with all the information collected from the existing facilities. These virtual tools allow the application of smart solutions that combine several urban configurations.

**Author Contributions:** Conceptualization, L.A., H.G., S.S., P.S.N. and J.A.F.; methodology, L.A., H.G., S.S., P.S.N. and J.A.F.; statistical study, S.S.; SWOT analysis, S.S. and L.A.; discussion, L.A., S.S. and P.S.N.; writing—original draft preparation, L.A., S.S. and P.S.N.; writing—review and editing, L.A., S.S. and P.S.N. All authors have read and agreed to the published version of the manuscript.

**Funding:** This research received no external funding.

**Acknowledgments:** This research has been developed in the framework of module 2 of the EERA Joint Program on Smart Cities. Authors would like to acknowledge the support given by the members of the EERA Joint Program on Smart Cities who have submitted the factsheet.

**Conflicts of Interest:** The authors declare no conflict of interest.

## References

1. European Commission. *The Strategic Energy Technology Plan—At the Heart of Energy Research and Innovation in Europe*; Communication from the Commission: Brussels, Belgium, 2018; ISBN 978-92-79-77316-7.
2. United Nations. Department of Economic and Social Affairs, Population Division. In *World Urbanization Prospects: The 2018 Revision*; ST/ESA/SER.A/420; United Nations: New York, NY, USA, 2019; ISBN 978-92-1-148319-2.
3. Balaras, C.A.; Droutsa, K.G.; Dascalaki, E.; Kontoyiannidis, S.; Moro, A.; Bazzan, E. Urban Sustainability Audits and Ratings of the Built Environment. *Energies* **2019**, *12*, 4243. [CrossRef]
4. Sustainable Development—17 Goals to Transform Our World. United Nations. Available online: https://www.un.org/sustainabledevelopment/ (accessed on 30 September 2020).
5. European Commission. The European Green Deal, Volume 53. 2019. Available online: https://eur-lex.europa.eu/legal-content/EN/TXT/HTML/?uri=CELEX:52019DC0640&from=EN (accessed on 30 September 2020).
6. Fabbri, M.; DeGroote, M.; Rapf, O. *Building Renovation Passports: Customised Roadmaps towards Deep Renovation and Better Homes*; Buildings Performance Institute European: Brussels, Belgium, 2016; Volume 46, ISBN 9789491143175.
7. Cartalis, C.; Synodinou, A.; Proedrou, M.; Tsangrassoulis, A.; Santamouris, M. Modifications in energy demand in urban areas as a result of climate changes: An assessment for the southeast Mediterranean region. *Energy Convers. Manag.* **2001**, *42*, 1647–1656. [CrossRef]
8. Sánchez, M.; Soutullo, S.; Olmedo, R.; Bravo, D.; Castaño, S.; Jiménez, M. An experimental methodology to assess the climate impact on the energy performance of buildings: A ten-year evaluation in temperate and cold desert areas. *Appl. Energy* **2020**, *264*, 114730. [CrossRef]
9. Pacheco, R.; Ordonez, J.; Martinez, G. Energy efficient design of building: A review. *Renew. Sustain. Energy Rev.* **2012**, *16*, 3559–3573. [CrossRef]
10. Kheiri, F. A review on optimization methods applied in energy-efficient building geometry and envelope design. *Renew. Sustain. Energy Rev.* **2018**, *92*, 897–920. [CrossRef]
11. Cao, X.; Dai, X.; Liu, J. Building energy-consumption status worldwide and the state-of-the-art technologies for zero-energy buildings during the past decade. *Energy Build.* **2016**, *128*, 198–213. [CrossRef]
12. Rezaie, B.; Rosen, M.A. District heating and cooling: Review of technology and potential enhancements. *Appl. Energy* **2012**, *93*, 2–10. [CrossRef]
13. Taleghani, M.; Tenpierik, M.; Kurvers, S.S.R.; Dobbelsteen, A.V.D. A review into thermal comfort in buildings. *Renew. Sustain. Energy Rev.* **2013**, *26*, 201–215. [CrossRef]
14. Yang, W.; Moon, H.J. Combined effects of acoustic, thermal, and illumination conditions on the comfort of discrete senses and overall indoor environment. *Build. Environ.* **2019**, *148*, 623–633. [CrossRef]
15. Heibati, S.; Maref, W.; Saber, H.H. Assessing the Energy and Indoor Air Quality Performance for a Three-Story Building Using an Integrated Model, Part One: The Need for Integration. *Energies* **2019**, *12*, 4775. [CrossRef]

16. Reynders, G.; Lopes, R.A.; Marszal-Pomianowska, A.; Aelenei, D.; Martins, J.; Saelens, D. Energy flexible buildings: An evaluation of definitions and quantification methodologies applied to thermal storage. *Energy Build.* **2018**, *166*, 372–390. [CrossRef]
17. Eshraghi, A.; Salehi, G.; Heibati, S.; Lari, K. An enhanced operation model for energy storage system of a typical combined cool, heat and power based on demand response program: The application of mixed integer linear programming. *Build. Serv. Eng. Res. Technol.* **2018**, *40*, 47–74. [CrossRef]
18. Calise, F.; D'Accadia, M.D.; Piacentino, A. Exergetic and exergoeconomic analysis of a renewable polygeneration system and viability study for small isolated communities. *Energy* **2015**, *92*, 290–307. [CrossRef]
19. Hiremath, R.; Shikha, S.; Ravindranath, N. Decentralized energy planning; modeling and application—A review. *Renew. Sustain. Energy Rev.* **2007**, *11*, 729–752. [CrossRef]
20. Drysdale, D.; Jensen, L.K.; Mathiesen, B. Energy Vision Strategies for the EU Green New Deal: A Case Study of European Cities. *Energies* **2020**, *13*, 2194. [CrossRef]
21. Macedo, J.; Rodrigues, F.; Tavares, F. Urban sustainability mobility assessment: Indicators proposal. *Energy Procedia* **2017**, *134*, 731–740. [CrossRef]
22. Lindenau, M.; Böhler-Baedeker, S. Citizen and Stakeholder Involvement: A Precondition for Sustainable Urban Mobility. *Transp. Res. Procedia* **2014**, *4*, 347–360. [CrossRef]
23. Serna, A.; Gerrikagoitia, J.K.; Bernabé, U.; Ruiz, T. Sustainability analysis on Urban Mobility based on Social Media content. *Transp. Res. Procedia* **2017**, *24*, 1–8. [CrossRef]
24. European Environment Agency. Air Quality in Europe—2018 Report. Available online: https://www.eea.europa.eu/publications/air-quality-in-europe-2018 (accessed on 30 September 2020).
25. Morawska, L.; Thai, P.K.; Liu, X.; Asumadu-Sakyi, A.; Ayoko, G.; Bartonova, A.; Bedini, A.; Chai, F.; Christensen, B.; Dunbabin, M.; et al. Applications of low-cost sensing technologies for air quality monitoring and exposure assessment: How far have they gone? *Environ. Int.* **2018**, *116*, 286–299. [CrossRef]
26. Martínez-Bravo, M.M.; Martínez-del-Río, J.; Antolín-López, R. Trade-offs among urban sustainability, pollution and livabiity in European cities. *J. Clean. Prod.* **2019**, *224*, 651–660. [CrossRef]
27. Van Broekhoven, S.; Vernay, A.L. Integrating Functions for a Sustainable Urban System: A Review of Multifunctional Land Use and Circular Urban Metabolism. *Sustainability* **2018**, *10*, 1875. [CrossRef]
28. Ferreira, V.; Barreira, A.P.; Loures, L.; Antunes, D.; Panagopoulos, T. Stakeholders' Engagement on Nature-Based Solutions: A Systematic Literature Review. *Sustainability* **2020**, *12*, 640. [CrossRef]
29. Margerum, R.D. Evaluating Collaborative Planning:Implications from an Empirical Analysis of Growth Management. *J. Am. Plan. Assoc.* **2002**, *68*, 179–193. [CrossRef]
30. Concerto European Initiative. Available online: https://www.concertoplus.eu/ (accessed on 30 September 2020).
31. Horizon. 2020. Available online: https://ec.europa.eu/programmes/horizon2020/en (accessed on 30 September 2020).
32. SET-Plan Working Group. Europe to Become a Global Role Model in Integrated, Innovative Solutions for the Planning, Deployment, and Replication of Positive Energy Districts. SET-Plan Action No 32 Implement Plan, 1–72. 2018. Available online: https://setis.ec.europa.eu/system/files/setplan_smartcities_implementationplan.pdf (accessed on 30 September 2020).
33. Ferrara, M.; Monetti, V.; Fabrizio, E. Cost-Optimal Analysis for Nearly Zero Energy Buildings Design and Optimization: A Critical Review. *Energies* **2018**, *11*, 1478. [CrossRef]
34. Sartori, I.; Napolitano, A.; Voss, K. Net zero energy buildings: A consistent definition framework. *Energy Build.* **2012**, *48*, 220–232. [CrossRef]
35. Magrini, A.; Lentini, G.; Cuman, S.; Bodrato, A.; Marenco, L. From nearly zero energy buildings (NZEB) to positive energy buildings (PEB): The next challenge—The most recent European trends with some notes on the energy analysis of a forerunner PEB example. *Dev. Built Environ.* **2020**, *3*, 100019. [CrossRef]
36. Marique, A.-F.; Reiter, S. A simplified framework to assess the feasibility of zero-energy at the neighbourhood/community scale. *Energy Build.* **2014**, *82*, 114–122. [CrossRef]
37. Amaral, A.R.; Rodrigues, E.; Gaspar, A.R.; Gomes, Á. Review on performance aspects of nearly zero-energy districts. *Sustain. Cities Soc.* **2018**, *43*, 406–420. [CrossRef]
38. JPI Urban Europe. Available online: https://jpi-urbaneurope.eu/ (accessed on 30 September 2020).
39. European Energy Research Alliance (EERA). Available online: www.eera-set.eu/ (accessed on 30 September 2020).
40. EERA. Joint Programme on Smart Cities. Available online: www.eera-sc.eu/ (accessed on 30 September 2020).

41. Köppen, W.; Geiger, R. *Das Geographische System der Klimate. Handbuch der Klimatologie*; Bulletin of the American Geographical Society: Berlin, Germany, 1936; Volume 43. Available online: http://koeppen-geiger.vu-wien.ac.at/pdf/Koppen_1936.pdf (accessed on 30 September 2020).
42. Vetme University of Wien. Available online: http://koeppen-geiger.vu-wien.ac.at/ (accessed on 30 September 2020).

**Publisher's Note:** MDPI stays neutral with regard to jurisdictional claims in published maps and institutional affiliations.

*Article*

# Possibilities of Transition from Centralized Energy Systems to Distributed Energy Sources in Large Polish Cities

**Dorota Chwieduk *, Wojciech Bujalski and Bartosz Chwieduk**

Institute of Heat Engineering, Faculty of Power and Aeronautical Engineering, Warsaw University of Technology, 00665 Warsaw, Poland; wojciech.bujalski@pw.edu.pl (W.B.); bartosz.chwieduk@itc.pw.edu.pl (B.C.)
* Correspondence: dorota.chwieduk@itc.pw.edu.pl

Received: 1 October 2020; Accepted: 9 November 2020; Published: 17 November 2020

**Abstract:** The main aim of this paper is to evaluate the possible transition routes from the existing centralized energy systems in Polish cities to modern low-emission distributed energy systems based on locally available energy sources, mainly solar energy. To evaluate these possibilities, this paper first presents the current structure of energy grids and heating networks in Polish cities. A basic review of energy consumption in the building sector is given, with emphasis on residential buildings. This paper deals with the evaluation of the effectiveness of operation of central district heating systems and heat distribution systems; predicts the improvement in the effectiveness of the energy production, distribution, and use; and analyzes the possible integration of the existing system with distributed energy sources. The possibility of the introduction of photovoltaic (PV) systems to reduce energy consumption by residential buildings in a big city (Warsaw) is analyzed. It is assumed that some residential buildings, selected because of their good solar insolation conditions, can be equipped with new PV installations. Electricity produced by the PV systems can be used on site and/or transferred to the grid. PV energy can be used not only for lighting and electrical appliances in homes but also to drive micro- and small-scale heat pumps. It is assumed that the PV modules are located on roofs of residential buildings and are treated as individual micro scale energy systems of installed capacity not larger than 50 kW for each of the buildings. In such a case, the micro energy system can use the grid as a virtual electricity store of 70% or 80% efficiency and can produce and transfer electricity using a net-metering scheme. The results show that the application of micro-scale PV systems would help residential buildings to be more energy efficient, reduce energy consumption based on fossil fuels significantly, and even if the grid cannot be used as a virtual electricity store then the direct self-consumption of buildings can reduce their energy consumption by 30% on average. Development of micro-scale PV systems seems to be one of the most efficient options for a quick transformation of the centralized energy system in large Polish cities to a distributed energy one based on individual renewable energy sources.

**Keywords:** centralized energy systems; district heating systems; energy demand in buildings; distributed energy sources; photovoltaic systems

---

## 1. Introduction

The idea of smart cities is not new for Poland. However, the concept of smart cities has been developed mainly from the standpoint of the global aspects of the smart management of a city. The global key indicators are as follows: the efficient administration of the city, cooperation between the city (administrative bodies) and inhabitants, service availability, transport access, social life, clean environment, and land use. The focus is put on social aspects to assure good standards of living and comfort, combined with a sense of security in everyday life as a citizen. Energy aspects such as energy efficiency, the safety of energy supply, and the utilization of renewables are also considered, but less stress is put on them. Such a situation has been caused by good access to energy networks, mainly centralized heating networks and grids, in most of the cities since socialist times. However, nowadays, society—especially young people—is becoming more interested in climate change issues, and therefore energy availability and the efficient use of energy has become one of the most important issues for sustainable development.

During the transformation from the centrally planned to free market economy, at the end of the last century (in the 1990s of the XX century) the restructuring and privatization processes of the Polish centralized economy, including the energy sector, started. The energy power sector was segmented into generation, transmission, and distribution. Competition among independent companies in each sector was introduced at the end of the 1990s. The central heat supply used to play and still plays an important role in the Polish energy economy. Heat is mainly generated in Combined Heat and Power (CHP) stations and in heat plants (which used to be public/state-owned). The fuel structure of direct consumption has been changing very slowly. Large cities and factories are supplied by coal-fired Combined Heat and Power stations. New trends can be seen as the solid fuel share (hard and brown coal) reduces in favor of liquid and gaseous fuels. The share of renewable energy is still very small. Up to now, district heating has been a major component of Poland's energy infrastructure. District Heating Networks (DHN) supply more than half of Poland's residential heating needs, with 75% in urban areas. The largest district heating network in Europe (and second in the world) is in Warsaw. It is noticeable that, since the last century, the heating demand has been decreasing continuously, mainly because of the intensive thermal refurbishment of buildings. However, electricity demand is growing. The reasons for this are mainly connected with an observed tendency for rapid increase in the penetration of electric appliances in the tertiary sector as well as in households.

Taking into account the existing situation in big cities, where energy systems and urban networks are very complex and most of them are centralized, it seems very difficult to assure the efficient transition from centralized energy systems to distributed energy sources in large Polish cities. Now, it is almost impossible to switch off the cogeneration power plants located in big cities, but it is necessary to modernize them to create low-emission plants, mainly through switching from coal to another (low emission) fuel. It is also not possible to cease using the central district heating networks, but it is possible to make them more efficient (e.g., through reduction in heat losses) and connect them not only with the central cogeneration plants but also with modern low-emission distributed energy systems based on locally available energy sources, such as renewables or waste heat (e.g., from industry or the tertiary sector). However, this can appear rather difficult, because central district heating systems are high-temperature networks (in Poland). When distributed energy sources such as renewables or waste heat sources are available, they supply heat at a medium or low temperature level. Therefore, when such sources can be utilized it is good to consider the possibility of the development of distributed local heating systems and distributed local electrical grids, not connected to large centralized energy systems but connected to each other. In such a way, low-temperature distributed local heating systems based on different renewable energy sources or waste heat sources can be developed to create local small- or even micro-scale centralized heating systems.

The transition from centralized mono-energy (based on coal) systems to integrated multi-energy source systems represents a great challenge, but it is a necessity in the present day. In countries like Poland, after so many years of constructing and using central high-temperature heating networks, which are used to supply heat to nearly every part of the city, it seems difficult or even impossible to switch to low-temperature distributed heating networks based on renewables in a short time. What is more, the heat is generated in cogeneration heating plants and such plants is promoted by EU energy policy [1]. Therefore, this process should be organized not very rapidly, rather than step by step, starting from small city areas (e.g., for a district of a few new buildings), which can be termed new housing estates of zero-net energy consumption, and then in a further step including positive energy districts.

Due to the existence of a central power grid and centralized district heating systems, the integrated management of energy in cities is a well-known process. However, it is not easy to determine the real reduction in energy consumption when the energy system is based on coal-fired plants. The transition of the centralized energy economy to a modern circular economy based on renewables is something completely new for the Polish energy sector at large and medium scales. To present the background of the present energy situation in cities, this paper evaluates the effectiveness of the operation of central district heating systems and heat distribution systems to predict the improvement in the effectiveness of energy production, distribution, and use. Then, it analyzes the energy effectiveness of the existing CHP plants in Warsaw. This paper evaluates the modernization possibilities of the existing central district heating systems in terms of reducing primary energy consumption and greenhouse emissions. A broad analysis of the possibilities of upgrading and developing district heating systems in Poland was presented by Wojdyga [2]. His paper deals with the history of the development of Polish heating systems and the current conditions. The diagnosis from 4 years ago states that district heating systems will continue to develop, but the development of the complementary elements in the form of dispersed small energy systems will be necessary.

It should be pointed out that the problem of the modernization of the centralized energy sector and especially of district heating systems has been problematic for many past socialist countries, including Poland. An example of the transformation of traditional district heating systems in Chemnitz (Germany) has been presented by Urbaneck and co-authors [3]. In the paper, the authors analyzed the possibility of separating a part of the heating network and switching it to a low-temperature section of the heating network with the use of a solar plant. In the conclusions, they state that the integration of solar heat requires special concepts. This applies in particular to areas with conventional districts. This applies in particular to areas with conventional district heating. Competition between solar heat and cogeneration should be avoided. Lygnerud discussed the situation of the district heating in Brasov (Romania) [4]. Inefficient infrastructure and the problem of loss of consumers due to unreliability of supply over previous decades were observed. The assessment of the impact of different policies on the feasibility of renewable energies and efficient heating was performed. The possibility of transformation into an efficient system was presented. The analysis showed what kind of politics should be applied. Problems of post-communist district heating systems were also discussed in [5]. The paper focused on the legal aspects of this kind of district heating system. Nowadays, Polish cities are expanding into new areas–country land (mainly agricultural in origin). The way of life is changing. People want to work in a city, but they want to live outside it, in new clean suburbs having some of privacy compared with high-rise multi-apartments housing estates. The new suburbs are developing very quickly.

Very often, such new suburbs do not have access to the central district heating system and other energy media such as gas, and sometimes they are even quite far from the power grid. Such a situation gives a unique opportunity for the introduction of distributed energy systems based on renewables for the generation of heat and electricity. In this way, the new districts of a city can be based fully on their own energy sources, where renewables can play the most important role.

A new trend, the electrification of the heating sector, is also evident in Poland and can be effectively applied in new districts. Electrification in this new sense means the application of electrically driven heat pumps for the hot water and space heating and using the heat pumps in a reverse cycle for cooling demand, if applicable. Heat pumps can be based on renewable energy heat sources (ground, water, air) or waste heat available at the location. Electrical energy supplied to heat pumps can be generated by local photovoltaic or wind turbines systems located on buildings or in their vicinity. On a larger scale, such districts can be supplied by the biomass or biogas CHP plants.

In most European countries, the concept of distributed energy systems in cities has been developed mainly towards the modernization of existing heating systems and extending them to new territory, but in the form of low-temperature heating systems based on renewables [6–8]. Most of such low-temperature heating systems are supplied with heat generated in solar thermal plants [8,9]. Solar thermal plants as distributed energy systems are a very promoting solution for low-temperature district heating systems in most western European countries. However, coupling them to high-temperature districts heating systems as is common in Eastern European countries is not yet possible because of technological but mainly economic reasons [3]. The energy sector and especially the district heating systems in the large cities of past socialist countries are completely different to the systems in other European countries. The path towards the deployment of distributed energy systems based on renewables must be based on energy roadmaps developed especially for different regions of Europe, even if there is one common EU target to reach the 80% reduction in annual greenhouse gas emissions in 2050 compared to the 1990 levels [7]. Sustainable energy generation and supply is expected through moving to decentralized and smart energy grid solutions [10]. The paper shows one of the possible routes for East European countries to reach the goal of sustainability.

This paper deals with the problem of the possible transition from the existing centralized energy systems in Polish cities to modern low emission distributed energy systems based on locally available energy sources, mainly solar energy. To evaluate these possibilities, the paper first in Section 2 presents the schematic idea of the existing energy structure in large cities and their suburbs. In Section 3, the current structure of the energy sector—mainly power and heating networks—in Polish cities is presented. Then, in Section 4 a basic review of the energy consumption in the building sector is given, with emphasis put on residential buildings. This is due to the fact that the hypothetical new low-energy district in the suburbs of the city proposed for consideration here is composed mainly of residential buildings. Such a new suburb housing estate and its distributed energy system based on renewables is presented in Section 5. The effectiveness of operation of the proposed energy system is analyzed, and the expected reduction in energy consumption is evaluated. The last chapter presents the conclusions of the studies conducted and some recommendations for transition steps to achieve low-energy or even positive-energy districts in large cities previously with a monopoly of large centralized energy systems.

## 2. Schematic Picture of the Energy Sector in Large Polish Cities

Nowadays, the phenomenon of cities spilling into rural areas can be seen in many countries. Such a situation is also typical for large Polish cities and their new suburbs, which can be called new green residential districts. People want to move out of the busy, noisy, and polluted cities to live in a clean and quiet habitat. It is much easier to assure low energy consumption in such new districts because it is easier to build new buildings with new energy sources according to the energy efficiency and clean environment standards than to transform (refurbish) old highly energy-intensive buildings supplied by inefficient centralized energy systems constructed on the basis of past technology.

The paper tries to show how difficult it is to transit the energy economy from centralized mono-energy fossil fuels systems, mainly based on coal, to distributed energy systems based on renewables. To give the basis for discussion, it is necessary to present the situation in the energy sector in large cities in a schematic way. Figure 1 presents such a schematic picture of the centralized energy sector typical for large Polish cities and its influence on other main sectors. This picture also shows what is happening in the new suburbs of these cities. Such new urban districts are developing quickly and they are supplied by their own energy systems. This is an indication of how cities are evolving.

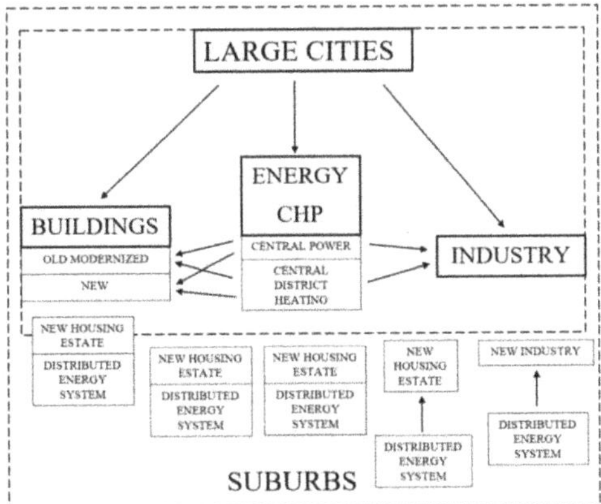


**Figure 1.** A schematic picture of the centralized energy sector typical for large Polish cities as well as its influence on other sectors and the city's suburbs with new housing estates coupled with distributed energy systems.

Figure 1 gives a basis for consideration and explains why specific sectors are analyzed in the following chapters of the paper. As can be seen in the figure, in large cities buildings depend completely on the energy sector. There is an impact of energy plants and networks on buildings, but not the opposite. In new districts in suburbs, there is mutual interaction between houses and energy systems. This is especially true when buildings are equipped with energy devices and installations which generate energy just for the building themselves. What is more, the elements of such systems can be installed using building integrated technologies (e.g., BIPV—building integrated photovoltaics).

The transport sector is not included in Figure 1 and in consideration in the paper. This is due to the fact that the present energy sector in large cities is not connected in a direct or indirect way with the transport sector. However, the transformation of the transport sector towards electrical vehicles is becoming a reality. It can be expected that electrical cars driven by renewable energy will be one of the important elements of future energy systems in cities. Such vehicles will be present in new low-energy city districts in the suburbs of the cities, acting as stores of electricity gained by micro renewable energy systems such as photovoltaic and wind systems.

## 3. Polish Energy Sector

### 3.1. District Heating Systems

The current situation of Polish district heating systems with a focus on heating networks in Polish cities is presented in this chapter.

Polish district heating systems have a long history. The biggest development of Polish district heating systems was carried out in the sixties and seventies of the last century. In those years, a very large development of heat demand was planned. In the 1980s and early 1990s, there was a significant decline in the development of industry and construction in Poland. As a result, the dynamics of the increase in heat demand decreased. In the late nineties of the last century, there was a great wave of thermal modernization of buildings. This caused the capacity of the systems to be much greater than the heat demand. Due to the excess of installed capacity, it was not necessary to modernize the heating systems. The history of development and decisions made over time has influenced the current structure of heating systems. Political decisions about the development of the DHN (District Heating Network) caused almost all cities in Poland to have heating systems. Poland is a leading supplier of heat for district heating networks [11]. The total district heat sales are presented in Figure 2.

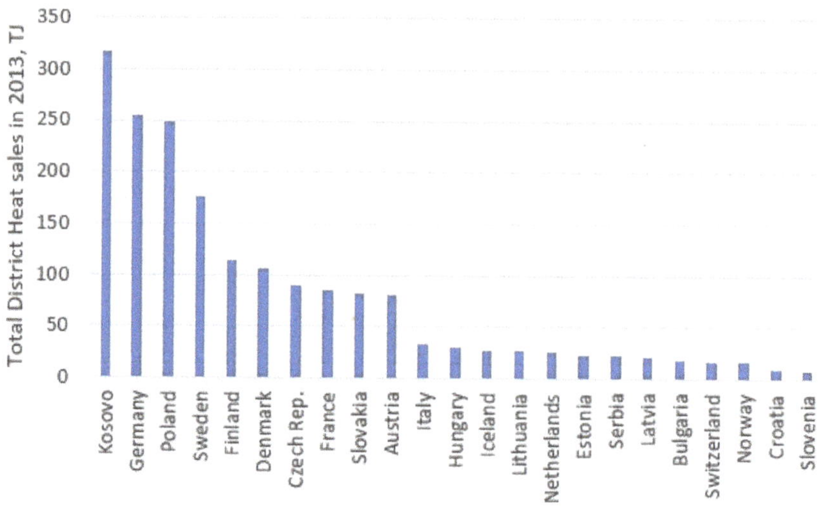


**Figure 2.** Total district heat sales in 2013 in European countries.

The great potential of the district heating systems in Poland is an important advantage. The same situation caused there to be a very unfavorable fuel structure for heat production for DHN. The fuel structure of the heat production for district heating systems with a capacity above 5 MW (in fuel) is shown in Figure 3 [12]. The presented data show that over 70% of the heat for DHN is produced with the use of hard coal, and the heat from renewable energy sources—i.e., biomass—is below 10%.

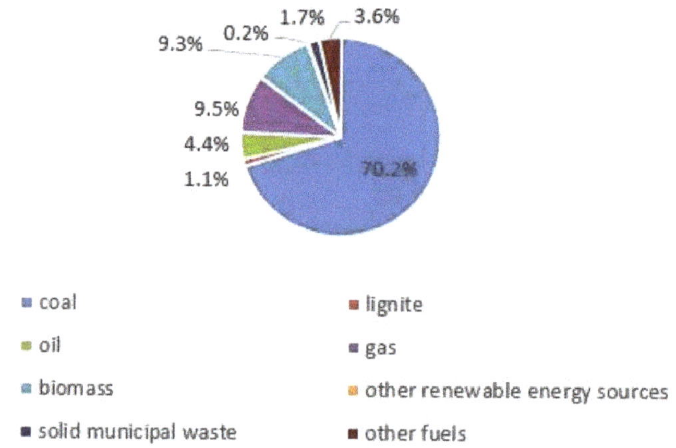


**Figure 3.** Fuel consumption for heat production.

The effectiveness of the system is very low because of the structure of the technology used in the systems. In the large power plants (from 100 MW to more the 2000 MW of thermal capacity), there are steam pulverized-fuel boilers with steam turbines. In the typical structure of medium and small systems, there are water stocker-fired boilers (coal-fired). The average efficiency of the transformation of prime energy to the used form of energy is presented in Figure 4. The specific emissions for unit of the heat are presented in Figures 5 and 6 [12]. In Figure 4, we can see that it is no significant progress in the efficiency of the Polish district heating systems. However, we can observe a decrease in the emissions of the $SO_2$ as a result of the installing desulphurization.

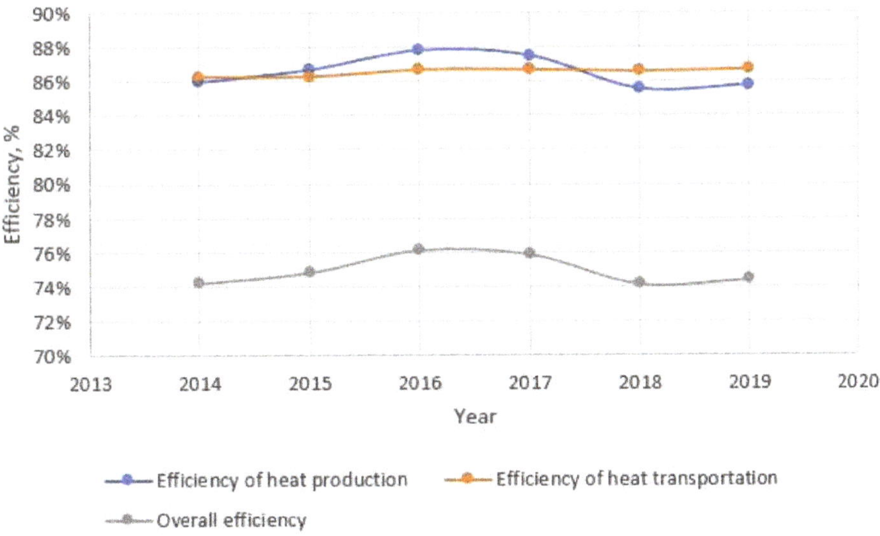


**Figure 4.** The efficiency of heat production and heat transportation—an average value for all Polish district heating companies over the last 6 years.

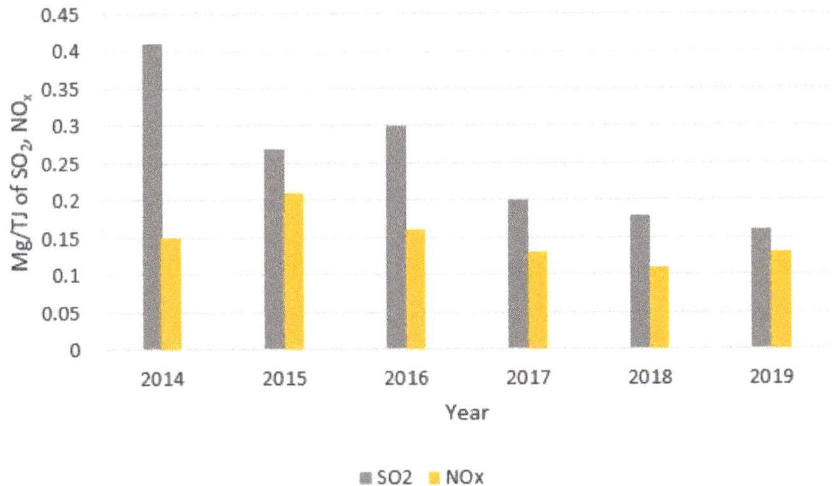


**Figure 5.** Average emission factors of $SO_2$ and $NO_X$ per unit of heat over the last few years.

**Figure 6.** Average emission factors of $CO_2$ per unit of heat over the last few years.

A comprehensive description of the Polish district system must be preceded by an analysis of the structure of CHP, non-CHP, and renewables in heat generation in Poland. Figure 7 illustrates the structure of heat production in Polish district heating companies. Most of the heat (57%) is generated by CHP units, because the large-capacity CHP units, which deliver heat to DHN, are the most suitable heat source for heat generation in Poland. Medium- and small-capacity plants are facing problems due to the lack of CHP units or renewables.

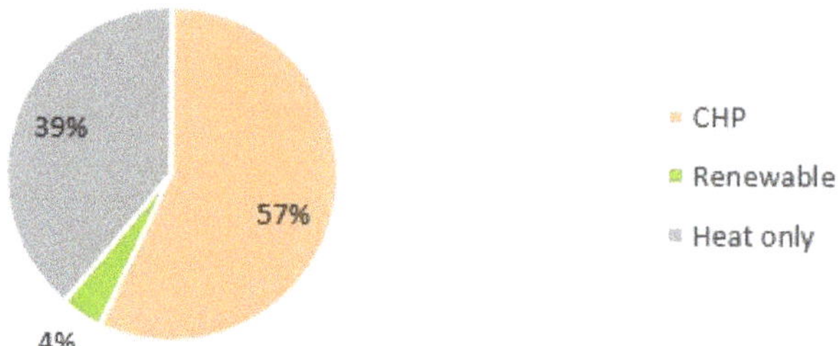


**Figure 7.** Structure of heat production in heating systems with a capacity greater than 5 MW in 2018—estimates based on data from the Energy Regulatory Office.

The indicators of the development of district heating systems is a debatable issue due to the lack of common methods for their definition. Authors suggest using the length of the district heating systems as a reliable indicator, because the length of DHN reflects the development of the district heating systems. The change in the overall length of Polish district heating systems over the last few years is depicted in Figure 8 [12]. The constant increase in the length of the district heating systems is observed over many years—i.e., 1% per year.

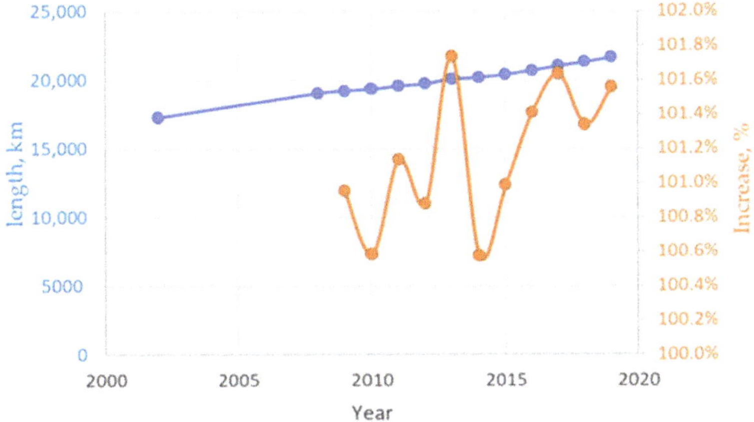


**Figure 8.** Overall length of the district heating network over the years.

Up to this moment, it seems that the overall condition of the Polish district heating system is good, however scrutiny of its efficiency and pollutant emissions reveals the need to change the heat production portfolio.

*3.2. Power Sector*

A similar situation applies to the power sector, which has not been significantly modernized for many years. Figure 9 presents the structure of electricity production in the years 1990 to 2019 [13]. It can be observed that over 90% of electric energy was generated by thermal power sources, while most of them are coal-fired power plants. In the years 2010 to 2015, an increase in electricity generation from renewable sources (i.e., wind power plants) is noticed. However, as a result of the change in legal regulations in 2016 (limitation of the height of the wind power plant depending on the distance from the buildings), this growth was stopped and currently the share of renewable energy is still very low.

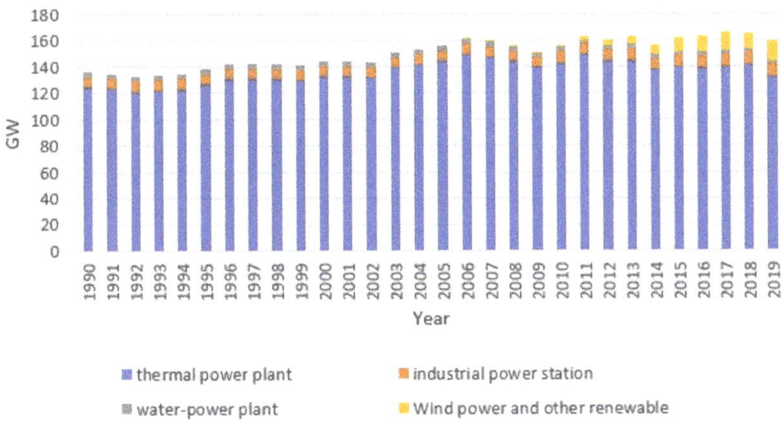


**Figure 9.** Electricity production in Poland in the years from 1990 to 2019 (GWh).

Recently, a rapid growth in the installed capacity in photovoltaic systems has been observed, which was caused by the introduction of a governmental support program. Figure 10 shows the increase in the installed PV capacity in recent years [14].

**Figure 10.** Installed capacity in photovoltaic in Poland in the years from 2010 to 2019.

On the other hand, dynamic changes in the way electricity is used are observed. The main change is a significant increase in power demand during the summer period, which results from an increase in air conditioning needs. The change in the value of the monthly average power demand in the last 10 years is shown in Figure 11 [13].

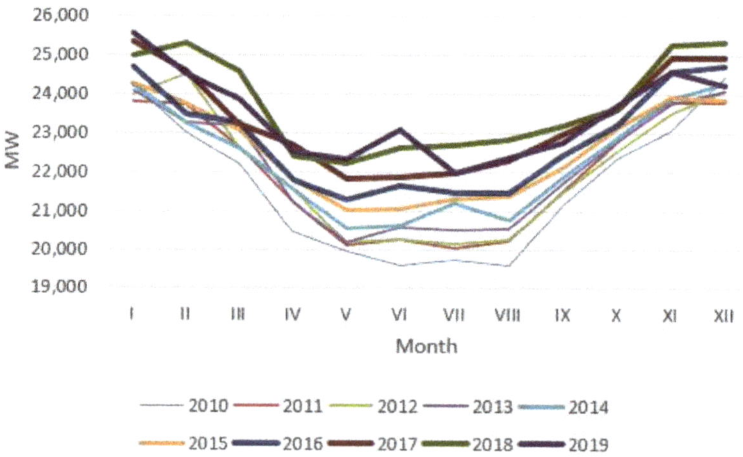


**Figure 11.** Average monthly domestic power demand in daily load peaks on working days from 2010 to 2019 (MW).

The discussed data shows that the Polish power system is based on outdated and inefficient technologies. On the other hand, the environmental forces are changing. Warsaw seems to be a good example. In Warsaw, most of the buildings are connected to the high-temperature district heating network, which is supplied by heat sources (in the year 2020, coal-fired CHP). Those CHP units are an important element of the city's electricity supply. It is obvious that, in the future, the district heating system will have to undergo transformation to fourth [15] and fifth generation systems [16]. The development of heat storage will be an essential element of the system transformation [17].

For technical reasons, it will be difficult to change the way the power is supplied. On the other hand, cities are developing intensively and new areas are being created which are located far beyond the network's coverage and are strongly urbanized. This creates a place for the intensive development of smart energy regions and positive energy districts (PEDs) in the future.

## 4. Building Sector in Poland

### 4.1. Old Buildings

As described in the previous chapter, changes in the energy sector are progressing slowly and are mainly concentrated on the modernization of the large coal-fired power plants. The power grids and district heating systems still need major upgrading. Much more is going on in buildings and the reduction in energy consumption is noticeable, as is presented in this chapter.

A basic review of the energy consumption in the building sector is presented at the beginning of this chapter, with the stress placed on residential buildings.

In Poland, as in many EU countries, the building sector is responsible for around 40% of the final energy consumption. The high heat demand in existing buildings is caused by past (mainly socialist) energy policy. In those days, energy prices were very low and there was no need to save energy. As a result, nobody was thinking about reduction in energy needs and energy consumption. The previous codes for buildings took into consideration only the heat transfer coefficients for walls and windows. As a result, the energy consumed in buildings was huge. In addition, the energy supplied to buildings was generated in inefficient plants fired by coal. Such a situation caused a high consumption of primary energy, as is presented in Table 1.

**Table 1.** Changes in the heat transfer coefficients for walls and windows and the indices of the final and primary energy consumption for heating energy (for space heating only) in the years 1966 to 2014.

| Time of Construction | Heat Transfer Coefficients for Walls and Windows [W/m$^2$K] | Final Energy Needs for Space Heating [W/m$^2$K] | Primary Energy Needs for Space Heating [W/m$^2$K] |
|---|---|---|---|
| –1966 | 1.16/–<br>1.40 */– | 240–280<br>300–350 | 364–447<br>489–569 |
| 1967–1985 (1983 $^{w}$) | 1.16/–(2.6) $^{w}$ | 240–280 | 364–447 |
| 1986–1992 | 0.75/2.6 | 160–200 | 244–325 |
| 1993–2001 (1998 $^{w}$) | 0.55/2.6 (2–2.6) $^{w2}$ | 120–160 | 203–244 |
| 2002–2008 | 0.30/2–2.6 | 90–120 | 69–183 |
| 2009–2013 | 0.30/1.8–1.7 | 90–120 ** | 69–183 ** |

* Two values of the heat transfer coefficients connected with a type of the building construction; $^{w}$ for windows, the limits for heat transfer coefficients were introduced for the first time in 1982 by the national standards; $^{w2}$ for windows, the range of limits for heat transfer coefficients were introduced in 1998 by the national regulations for buildings. ** Indices for the final and primary energy used to be dependent on the so-called "shape coefficient" of a building, which was the building surface area to volume ratio of a building.

It should be underlined that only the limits for the heat transfer coefficients for walls in the 1960s to 1980s were regulated by the official codes. For windows, official regulations were introduced in the year 1992. Extremely high indices of the final energy needs for space heating were caused by the very high values of these coefficients and very high air ventilation rates (very poor leak tightness of building envelopes). The high indices of primary energy needs resulted in using low-efficiency coal-firing in the heat and power (or CHP) plants. It is important to notice that the indices presented in Table 1 present the values of energy needs rather than of energy consumption, as are commonly used nowadays. In reality, in those days it was not possible to supply such a large amount of heat to buildings to assure internal thermal comfort. As a result, the thermal comfort was very poor. The indoor air temperature used to be much lower than that required for thermal comfort and living comfort. The indoor air humidity was high and condensation problems were very bad. As a result of the condensation, mold and fungus appeared. When, at the turn of the century, the transformation of the national economy started, most of the buildings constructed in the 1960s to 1990s went through an intensive thermal refurbishment process. As a result, since those days the energy demand of Polish buildings has decreased significantly. However, they still consume much more energy than new buildings. It is much easier to introduce modern energy systems in buildings where energy effectiveness has been introduced from the beginning (design and construction phase), and thanks to that they need less energy for running all their energy systems.

### 4.2. New Buildings

According to the EC directive of the energy performance of buildings [18,19], it is necessary to reduce energy consumption significantly and to achieve nearly zero-energy buildings by the beginning of 2021. Polish regulations set the limit of primary energy consumption indices for domestic hot water (DHW) and space heating for residential and non-residential buildings. Since the year 2021, multifamily buildings and single-family houses cannot consume more primary energy than 65 and 70 kWh/m$^2$ per year, respectively [20]. However, there are no restrictions on the final energy consumption. Table 2 presents all the recent changes (since the year 2014) in the official regulations for the indices of primary energy consumption for heating energy (space heating + DHW) and the heat transfer coefficients for walls and windows.

**Table 2.** Changes in the heat transfer coefficients for walls and windows and the indices of primary energy consumption for heating energy since the year 2014.

| Type of a Building and Duration of the Regulation | Heat Transfer Coefficients for Walls [W/m$^2$K] | Heat Transfer Coefficients for Windows [W/m$^2$K] | Indices of Primary Energy Consumption for Heating Energy, [kWh/m$^2$] |
|---|---|---|---|
| 2014–2016 | | | |
| multifamily buildings<br>single-family houses | 0.3 | 1.3 | 105<br>120 |
| 2017–2020 | | | |
| multifamily buildings<br>single-family houses | 0.23 | 1.1 | 95<br>85 |
| since 2021 | | | |
| multifamily buildings<br>single-family houses | 0.2 | 0.9–1.1 | 70<br>65 |

There are no official regulations for the limits of electrical energy consumed by electrical appliances and lighting and for cooling energy. In previous decades, because of the climate and poor thermal quality of residential buildings, there was no problem with the overheating of buildings. Even when heat was gained quickly during the day in summer due to high solar radiation, it was also lost very quickly because of the low thermal insulation quality (thin layers of insulation without the proper technology for attaching it to the building structure) of buildings' envelopes (high heat-transfer coefficients). Nowadays, when buildings are very well insulated and well-constructed, the heat cannot leave (be lost) so easily and quickly in a natural way, and a cooling system to remove the gained excess heat (excess quantity) has to be used to keep the indoor thermal comfort at the required level in summer [21]. Table 2 presents the changes in the official regulations for the indices of primary energy consumption for heating energy and the heat transfer coefficients for walls and windows, respectively.

Analyzing the data in Table 2, it can be seen that the present building codes require a high thermal quality of buildings and that, as a result, the heat transfer through envelopes is very much reduced. As a consequence, the heat demand for space heating can be very small even if the winters are quite severe.

It can be added that there are no limits for electricity consumption by electrical appliances and lighting. However, people have started reducing electrical energy consumption mainly because of economic reasons. The prices of electricity have started to increase significantly in recent years.

As mentioned regarding the new hypothetical housing estates considered in this paper, all buildings are constructed according to the present building codes and regulations, which means they can be called nearly zero-energy buildings (from the Polish building sector perspective).

## 5. New Warsaw's Suburb Supplied by Distributed Solar Energy Sources

### 5.1. A Micro-Scale PV System as Distributed Energy Source in Polish Conditions

As mentioned, the new residential areas in the suburbs of large cities are characterized by relatively low energy needs and more often than not use their own energy systems. Heating is mainly based on gas boilers, which can use liquid gas from tanks, or fireplaces (wood fired) coupled with air ducts to distribute the heat gained inside a house. The electrical power grid is available nearly everywhere in the country, so it is used to supply energy to electrical appliances. Sometimes electricity is used for space heating, and this is mainly in cases when residents install their own micro-PV system. The utilization of PV systems at the micro-scale has become quite popular recently, mainly because of the financial incentives supported by generous national regulations and also as a result of the increasing awareness of the environment and climate change.

In this chapter, the possibility of the introduction of PV systems as distributed energy sources to reduce energy consumption by residential buildings located in the new green area of Warsaw's suburbs is analyzed. It is assumed that residential buildings in that area have good solar irradiation conditions (roofs have good exposure to solar radiation, there are no shading devices on the roof and no shading obstacles in the vicinity (trees, other buildings, etc.)) and the PV installations can be used in effective way. In recent years, PV technology has become relatively cheap in Poland. Installing micro-scale PV systems has also became a very simple solution, not only because the technical method of installing the PV modules is simple, but also because PV modules are widely available and, as mentioned above, legal regulations positively support the use of this technology [22]. Micro wind power plants are not a popular technology. However, in aiming to achieve nearly zero- or even positive-energy districts, perhaps it is good to take into account the possibility of installing such systems to be complementary to PV installation.

The electricity produced by PV systems can be used on site and/or transferred to the grid. PV energy can be used not only for lighting and electrical appliances in homes but also to drive micro- and small-scale heat pumps. The potential for heat pump deployment is considered taking into account the existing district heating connections, which are located in the cities, but not in the newly emerging housing estates in the suburbs. The operation of heat pumps in a reversed cycle to supply cooling energy can be also applied (but is not analyzed in the paper). The approach is based on the existing regulation and support scheme for the application of renewables. Analysis is performed on the basis of calculations determining the energy needs of the considered buildings and the energy gains and effectiveness of energy systems. It is assumed that the PV installations are located on roofs of residential buildings and are treated as individual micro-scale energy systems of an installed capacity not larger than 50 kW for each of the group of buildings. Polish regulations put this capacity limit on micro-scale energy systems based on renewables. In such cases, the micro energy system can use the grid as a virtual electricity store of 80% (up to 10 kW of installed capacity) or 70% (between 10 and 50 kW of installed capacity) efficiency, and can produce and transfer electricity on the basis of a net-metering scheme [22]. Treating the grid as a store means that the energy not used in a building at the time when it is produced (there is no so called self-consumption) can be sent to the grid, and later when the PV system is not operating 80% or 70% of the energy transferred to the grid is taken out of the grid when it is needed. The maximum storage time is half a year, but with strictly defined dates. A year is divided into two halves, with the first starting on 1 January and the second on 1 July. However, there is official limitation of the storage capacity of the grid. The amount of energy transferred to the grid must be equal to energy consumed in a building which produces such energy. When more energy is transferred to the grid, it is treated as very cheap extra energy gained by the grid. The extra energy supplied to the grid can be sold by the energy producer to the energy distribution company (the owner of the grid), but the price for 1 kWh of such extra energy is at least ten times smaller than the regular price of electricity for individual consumers.

### 5.2. General Description of a Hypothetical Suburb Housing Estates

New single and multi-family housing estates are being built in the suburbs of large cities. They are built in rural areas. Such a location of a built housing estate is associated with certain limitations. As mentioned, such buildings usually do not have access to the central district heating or the gas network. At the initial stage of construction, the estates only have access to cold water from the mains and electricity from the grid. This creates the conditions to use renewable energy sources to cover the needs for domestic hot water and space heating. Figure 12 shows a typical new suburb housing estate.

**Figure 12.** A typical new housing estate in the suburbs of Warsaw.

This chapter describes a hypothetical estate of terraced single-family houses. A system with heat pumps has been proposed to ensure the needs for space heating and domestic hot water. The estate is supposed to produce the same or more energy than it consumes. This means that systems generating electricity should be installed in the buildings covering all available surfaces exposed to the solar radiation. Photovoltaic systems located on the roofs of houses can be used to power devices located in the buildings. In addition, small wind turbines can be located in the vicinity. The energy produced by renewable energy systems can be used by the household members or sent to the power grid.

First of all, basic assumptions about the location and orientation of the buildings of the housing estate under consideration have been made. As the estate is located in the suburbs of Warsaw, then meteorological data for this location are used for the calculation [23]. It has been assumed that the estate consists of 25 terraced houses divided into four buildings. Each of the houses has an area of 120 m$^2$ (of the heating space) and two floors. The schematic location of the buildings is shown in Figure 13.

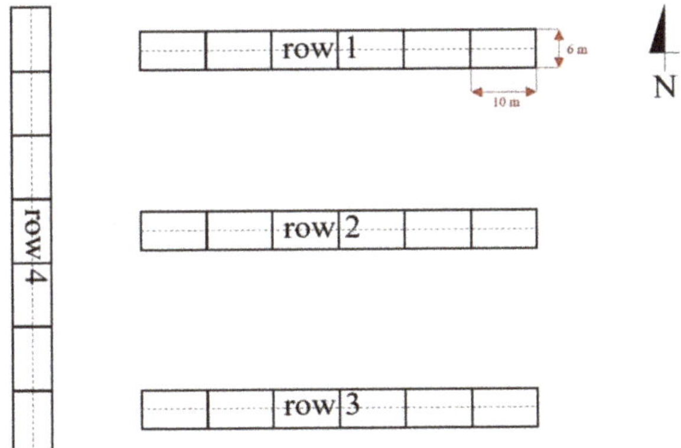


**Figure 13.** Location of buildings of the hypothetical estates.

Eighteen of the terraced houses have their roofs facing south/north and seven towards the east/west (this is one of the most typical arrangement for such estates). All the roofs are inclined at an angle of 30°. These are important data relevant to determining PV systems gains.

### 5.3. Methodology of Determination Energy Performance of Building and Energy Systems

To analyze the application of distributed energy systems based on renewables in the new suburbs of Warsaw, it was first necessary to conduct a study on energy needs of such a new housing district and then to select a renewable energy system to provide the energy needed. Next, the operation of the energy systems was modelled and simulated to determine the energy performance of the district. A special simulation code has been developed to simulate the building dynamics and renewable distributed energy systems.

The dynamics of the buildings have been considered taking into account changing weather conditions (ambient air temperature and solar irradiation). The energy balance of the building was formulated with all energy fluxes flowing out and in. Calculations of space heating (or cooling) needs are performed with a time step equal to one hour. The methodology of calculations is generally based on standards for the determination of energy performance of buildings [24]. The main difference between the standard method and model applied for the simulation is the time step of the calculations. The standard method uses average data for every month, while the model used in the study requires the hourly values of climatic data for the whole year. What is more, solar irradiation is calculated with regard to the specific orientation and inclination of a surface exposed to solar radiation using the Liu–Jordan isotropic diffuse solar radiation model [25]. To determine such solar irradiation, hourly values for direct and diffuse solar radiation on horizontal surface for given location are taken from the official data base [23]. Such a detailed model of the availability of solar radiation has been used because the solar radiation impact on the energy balance of buildings is especially high in new buildings that are very well insulated and airtight [21]. Solar radiation enters the interior of a building through transparent elements—windows—then is absorbed by the internal surfaces of a room, furniture, etc. As a result of the photo-thermal conversion of solar radiation, the indoor air temperature increases. Too much solar gain can cause the overheating of rooms.

The detailed calculations of solar irradiation are also important for calculations of energy gained by the photovoltaic systems. Operation of all energy systems is analyzed on hourly basis. Hourly values of wind speed in given location are taken from the official data base [23]. They are used for determination of energy generated by micro wind turbines. Operation of the heat pumps used for space heating and domestic hot water also changes in time depending on changing demands for both energy needs. As result performance of the heat pump also changes in time.

### 5.4. Energy Demand of a Hypothetical Suburb Housing Estates

To determine the electricity demand, it is necessary to assume an appropriate energy consumption profile. In the case of the estate in question, the total demand consists of two components. The first is the demand for energy to power household appliances and lighting. For the calculations, two different electricity consumption profiles were assumed depending on the lifestyle of the household members. The self-consumption of the energy produced by the PV system depends on the characteristics of electricity consumption in a given building. The self-consumption means that the electricity generated by the PV system at a given time is used by inhabitants at the same time. Figures 14 and 15 show the two proposed variants. Variant A represents the situation when some members of family are at home during the day. Variant B represents the situation when all members are out of the home (working, learning, etc.)

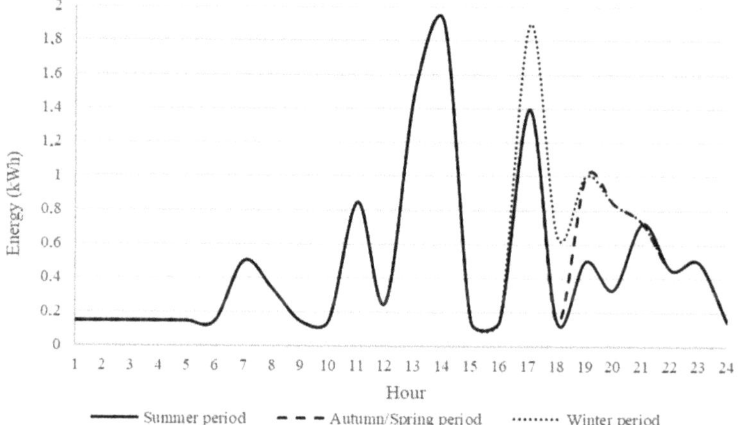


**Figure 14.** Electricity consumption profile (A).

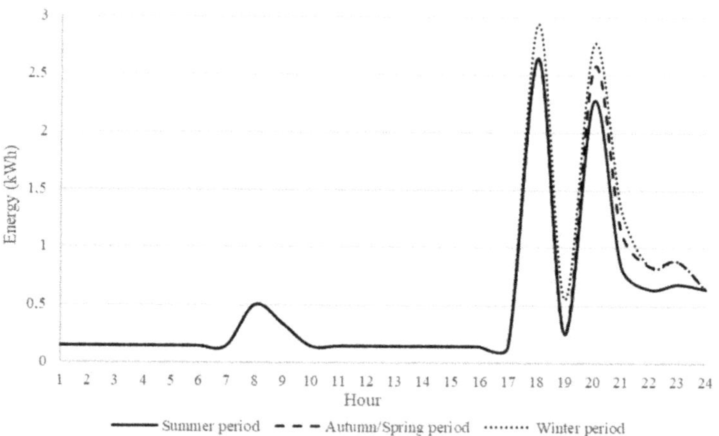


**Figure 15.** Electricity consumption profile (B).

It was assumed that 25% of the residents of the housing estate use energy according to profile A and 75% according to profile B. The daily electricity consumption in both cases is the same and amounts to approximately 12 kWh. The second factor is the energy required to power the heat pump.

In order to determine the demand for energy to power the heat pump, the demand for space heating, ventilation, and domestic hot water was determined with an hourly time step, as mentioned before. The demand for domestic hot water was calculated assuming that every household member consumes 40 L of domestic hot water of 50 °C temperature every day. The daily demand for domestic hot water for a family of four is approximately equal to 9.3 kWh. For domestic hot water consumption, two different water consumption profiles have also been taken into account. As the water is heated by a heat pump that uses electricity, the DHW consumption has an impact on the self-consumption of the energy generated by the PV system. Figures 16 and 17 present the assumed DHW consumption profiles, which are relevant to previous variants A (when some members of family are at home during daytime) and B (when all members are out of the house).

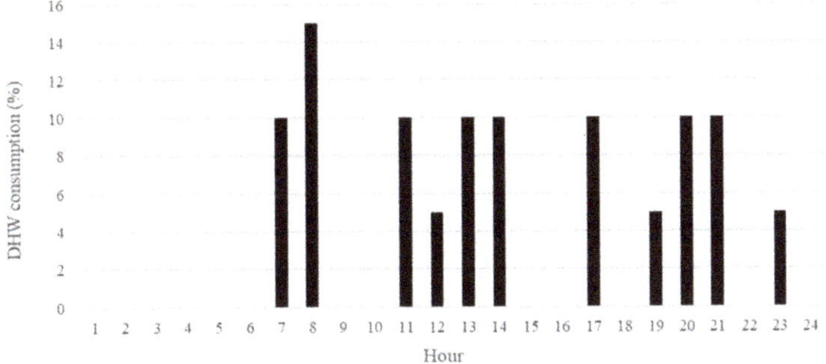


**Figure 16.** Domestic hot water consumption profile (A).

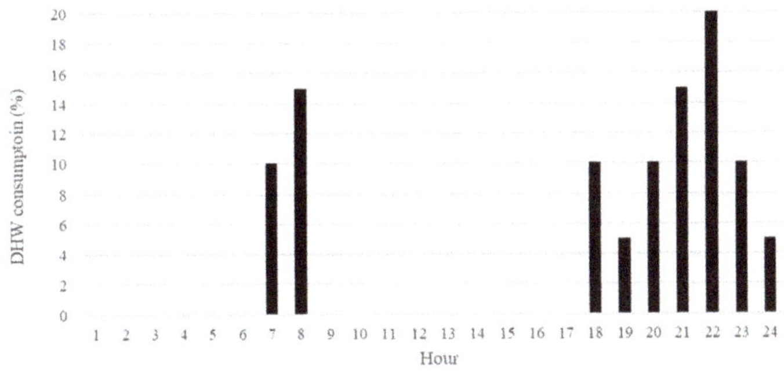


**Figure 17.** Domestic hot water consumption profile (B).

The demand for space heating has been determined as described in Section 5.3. The standard values of heat transfer *U* coefficients have been used according to the existing regulations published in "Technical Conditions for buildings" for the year 2021 [20]. This document contains the values of heat transfer coefficients that must be met by the individual building partitions. The heat transfer coefficient (*U* value) for external walls should be lower than 0.2 W/m$^2$K. The *U* coefficient for windows should be at a level of 0.9–1.1 W/m$^2$K. For the ground level floor, the *U* coefficient has been estimated to be equal to 0.17 W/m$^2$K. Knowing the dimensions of the individual partitions of a building, it was possible to determine the heat loss coefficient $H_{tr}$ W/K for such a building. Additionally, based on the volume of the ventilated space of a building and the air exchange rate, the value of the heat loss coefficient for ventilation $H_{ve}$ W/K was calculated.

The window area of each house accounts for 25% of the floor area—i.e., 24 m$^2$. The window area for the entire row of six houses is 144 m$^2$. The surface of the walls is 408, the roof is 403, and the floor on the ground is 360 m$^2$. The fourth row of houses inhabited by seven families has a correspondingly larger area of individual partitions. The $H_{tr}$ and $H_{ve}$ coefficients determined thanks to these data are 1530 and 1375 W/K. Based on these values, the power of the heat pump has been calculated. In the considered case, the heat pump supplies for the needs of space heating and domestic hot water; however, in the period of the highest energy demand much more energy is used to heat the building than to heat the domestic hot water. Therefore, when determining the heat pump capacity, the heat losses of the building are the most important factor. The total heat power of the heat pump appropriate for the housing estate should be about 70 kW.

Taking into account the location of the housing estate, it is possible to use four smaller heat pumps; one for each row of houses (i.e., three heat pumps of 17 kW of thermal power and one heat pump of 19 kW for the row extending from the south to the north direction). It can be added here that, apart from solar gains, the internal gains also have been taken into account and assumed to be at the level of 3 W/m$^2$ for the living space area [20]. On this basis, the heat balance was created for the entire estate, taking into account all components of energy losses and gains. Figure 18 shows the daily heat demand distribution for the space heating and ventilation.

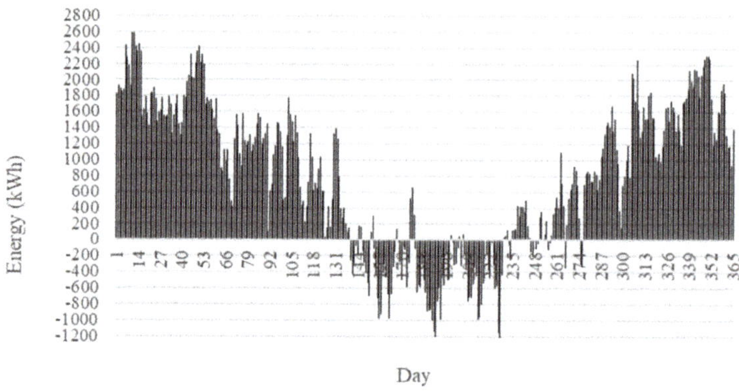


**Figure 18.** Daily heat demand for space heating and ventilation.

## 5.5. Operation of the Distributed Renewable Energy System

The calculations take into account that one of the rows of buildings faces a different direction than the others (see Figure 13). The total heat demand for heating and ventilation during the heating period is 315,500 kWh.

The previously presented assumptions for the calculations show that the approximate demand for electricity to power household appliances and lighting is 1.1 MWh per person per year. Assuming that the buildings are heated by ground brine-water heat pumps, the seasonal coefficient of performance (SCOP) was at the level of 5.3 (this is the averaged value of all calculated COPs). In the case of domestic hot water heating, the heat pump's SCOP is usually much lower [26]. The temperature to which the water should be heated is much higher than temperature of water required for underfloor heating systems. For domestic hot water heating, the SCOP coefficient is equal to 3. The annual heat demand for domestic hot water heating for the whole estate is approximately equal to 85,000 kWh. Using the above data and calculation results, the annual electricity demand of the housing estate has been determined. The annual demand for electricity in the discussed housing estate is approximately equal to 200 MWh. It turns out that the estate generates more electricity than it consumes; PV systems or small wind turbines installed in the estate should produce more than 200 MWh of energy. The wind conditions near Warsaw are not the best and allow the production of about 500 kWh of energy per year per 1 kW of rated power of the wind turbine. Figure 19 shows the average wind speeds in Warsaw in the year 2015 (Figure was elaborated using data presented in [27]).

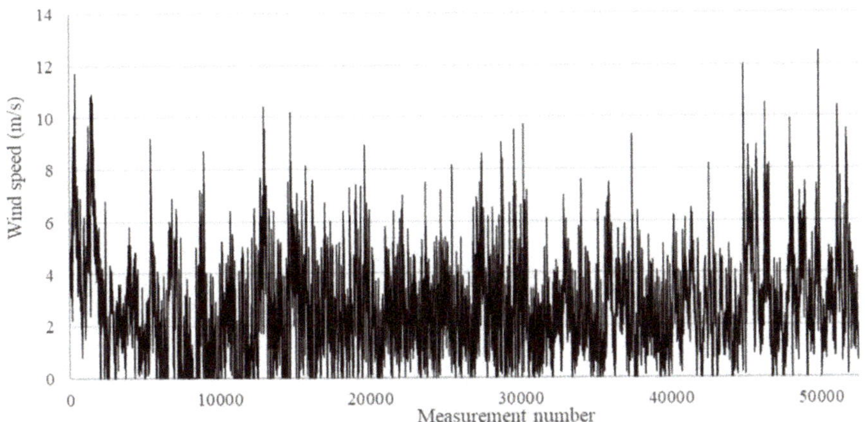


**Figure 19.** Average wind speed in Warsaw in the year 2015.

For this reason (low wind speed), the use of PV modules on the roofs of the housing estate has been considered as the priority. The results of the calculations show that the PV modules of the nominal power 1 $kW_p$ directed to the south and inclined at an angle of 30° can generate 1.15 MWh of energy per year. This means that such a system with a capacity of approximately 174 $kW_p$ can generate 200 MWh of energy per year. Currently, the most common photovoltaic modules have dimensions of 1 × 1.64 m and a power of 320 $W_p$. On the southern roofs of three rows of houses, 360 modules with a power of 115 $kW_p$ can be installed. This means that, in order for the estate to become positive energy, it is also necessary to install photovoltaic modules on the fourth building. The roof of this building faces east and west.

The annual energy gains from the installations installed on these sides is lower than in the southern side. The 115 $kW_p$ installation located on the southern roofs of the buildings generates approximately 130 MWh of energy. The systems on the fourth building must produce a total of 70 MWh of energy per year for the estate to be considered as zero or positive energy. In the Polish solar conditions, higher energy gains from PV installations are obtained from the west sides of buildings than the east (because of their better solar irradiation conditions [21]). Up to 140 modules can be installed on the west side. They will produce 38.7 MWh of energy annually. Due to the lower solar irradiation, a system of the same size located on the eastern side of the roof will produce 36 MWh of energy.

If all southern, eastern, and western roofs of buildings in the housing estate in question are fully covered with photovoltaic modules, the PV system has a power of 204.8 $kW_p$. It can generate about 207 MWh of energy throughout the year. The photovoltaic installation can be divided into five smaller systems, and then each of them has a power of less than 50 $kW_p$. Each roof slope can be a separate PV system. Systems with south-facing modules have a power of 38.4 $kW_p$. The installations located on the roof of the fourth building have a capacity of 44.8 $kW_p$. Figure 20 shows the total energy gains from all the considered PV systems.

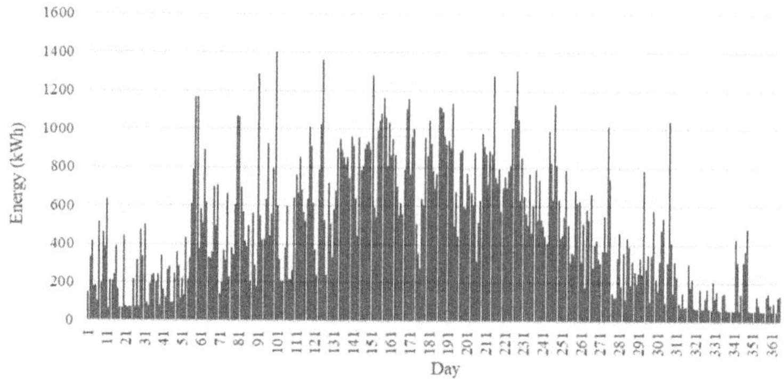


**Figure 20.** Energy gains from the photovoltaic (PV) system.

In order to summarize the results of the operation of the considered systems, Table 3 has been included. This table contains selected physical and energy parameters of the considered systems, as well as the annual energy demand for heat and electricity and the energy generated by the photovoltaic systems.

**Table 3.** Selected parameters and annual energy demand of the considered systems.

| | Building | | | | | |
|---|---|---|---|---|---|---|
| | 1 | 2 | 3 | 4 | | |
| | | | | East | West | Σ |
| PV system area [m$^2$] | 200 | 200 | 200 | 230 | 230 | 1060 |
| PV system capacity [kW$_p$] | 38.4 | 38.4 | 38.4 | 44.8 | 44.8 | 204.8 |
| PV system energy gains [MWh/year] | 44.1 | 44.1 | 44.1 | 36 | 387 | 207 |
| Electricity consumption [MWh/year] | 48.6 | 48.6 | 48.6 | 54.2 | | 200 |
| Heating demand [MWh/year] | 76.6 | 76.6 | 76.6 | 85.7 | | 315.5 |
| DHW demand [MWh/year] | 20.64 | 20.64 | 20.64 | 23.08 | | 85 |
| Heat pump capacity [kW] | 17 | 17 | 17 | 19 | | 70 |
| HP SCOP (heating) | - | - | - | - | | 5.3 |
| HP SCOP (DHW) | - | - | - | - | | 3 |

On the basis of these calculations, the self-consumption of the generated energy has been determined. For the energy consumption profile A, it is 35%, and for B it is 27%. The self-consumption for the entire estate can be approximately equal to 30%. The rest of the energy is transferred to the grid. When there is no energy generated by the PV system (low solar irradiance or nighttime), electricity is taken from the grid. The amount of energy sent to the grid by the PV system is higher than the energy taken by the housing estate from the grid.

## 6. Conclusions

One of the most important conclusions of the analysis of the energy sector in Poland is connected with the present situation of the district heating sector. This sector faces very great challenges. They result from the fact that for many years the heating sector (district heating systems) was practically not modernized. As such, there is currently a tremendous amount of work to do to transform it into a modern ecological heat source for consumers. In the opinion of the authors, this means that district heating systems will not develop territorially. Therefore, it is important to look for solutions for new areas of construction where the district heating systems will not reach. This paper presents a solution that seems to be technically simple, as well as energy efficient, giving a significant reduction in primary energy consumption that, moreover, it is in line with the latest trends in achieving energy self-sufficiency in new housing estates and sustainability in energy systems.

The paper has considered a hypothetical housing estate using ground source heat pumps for space heating and domestic hot water. It turns out that photovoltaic modules installed on the roofs of buildings can generate the same or even more energy than household members consume. The so-called self-consumption of generated energy (electricity) for the entire new housing estate in the suburbs of Warsaw can be at a level of 27–35% depending on the daily profile of energy consumption by the residents of houses. It gives on average of around 30% of total daily energy consumption. The results of the calculations show that, in Polish climatic conditions, it is possible to build housing estates based only on their own distributed energy sources, but nowadays they should be connected with the national power grid. When the self-generation of energy is not possible, then the energy from the grid is used. Most of the energy (70–80%) generated by the distributed energy systems cannot be used at the same time by residents, and energy is sent to the grid to be used by other end-users. It is evident that cooperation with the grid is necessary. The power grid is treated as a source of energy, a sink of energy, and an energy store. In such a case, it is not necessary to be connected to the central district heating system. Every building can be supplied by its own energy system, especially when heat pumps are applied. It can be concluded that, without the heat pumps and good cooperation between the PV systems and the power grid, it would not be possible to assure so effective an operation of the distributed renewable energy systems.

The next important conclusion is related to the buildings themselves. The new housing estates in the suburbs of large cities are designed and constructed according to new building standards and codes. This means that they need comparatively little energy to run their heating systems. In the case of single-family houses, there should be less than 65 kWh/m$^2$ (of living area) of annual primary energy used by the heating systems (see Table 2). To further reduce the primary energy consumption and make the new housing estates really independent of the central power grid, it is necessary for them to use their own distributed energy sources based on renewables. Nowadays, the utilization of solar energy is relatively simple and inexpensive. The conversion of solar radiation into useful thermal or electrical energy can be achieved thanks to mature and effective solar energy technologies. However, to assure the effective operation of such solar systems the solar irradiation conditions must be as good as possible. This means that the location of buildings in the new housing estates should be carefully chosen. All the area should be well exposed to the sun, with no shading obstacles in the surrounding such as trees and other buildings. An open area is recommended. Building houses in woods is not advisable (however, many new housing districts are located in such places). Roofs should be inclined (about 30°) and directed to the south to have enough surface for the installation of solar receivers (PV modules and solar thermal collectors). Buildings should be extended from east to west so that a sufficiently large area is exposed to solar radiation and it is possible to locate PV modules and solar thermal collectors there. Using solar energy in an active way makes the architectural and construction design of a building favorable for the use of solar energy in a passive way. As a consequence, the space heating demand is naturally reduced.

The results show that the application of micro-scale PV systems would help residential buildings to be more energy efficient and reduce energy consumption based on fossil fuels significantly. As a result, the buildings reduce their energy consumption by 30% on average in a direct way, and fully if the grid is treated as an energy store. The development of micro-scale PV systems seems to be one of the most efficient options for a quick transformation of the centralized energy system to a distributed energy one, based on individual renewable energy sources, in large Polish cities. In such a way, the traditional CHP plants and district heating networks located in the cities can be modernized step by step, giving room for actions to implement new investments in distributed energy sources based on renewables.

It can be mentioned that problem of transforming district heating systems (old systems) into modern energy systems is a very broad issue. Issues such as Demand Side Response (DSR), energy management and planning in cities, and control systems in buildings (e.g., Supervisory Control And Data Acquisition (SCADA) systems) are some of the top issues for this kind of research. However, this paper focusses on a much smaller problem. Two main aspects are analyzed: the great effort which has to be made to transfer the old district heating system to the modern one and an explanation of why it can be predicted that in Poland the district heating networks will not expand to the new areas. Cities will grow and new districts and housing estates have to be supplied with heat (and electricity, of course); that is why outside of the cities—in the cities' suburbs—another solution should be developed. This paper focusses on one of the possible solutions. This solution is for new housing estates to be supplied only by the electrical energy gained by PV systems and energy transferred from the national grid.

As a final conclusion, it can be stated that, paradoxically, the poor state of central energy systems, the complicated process of their modernization, and the lack of energy infrastructure in the suburbs of large cities favor the use of renewable energy sources in distributed energy systems in new housing estates constructed in remote, previously rural, areas.

**Author Contributions:** Conceptualization, D.C.; methodology, D.C. and B.C.; software, B.C.; validation, D.C. and W.B.; formal analysis, D.C. and W.B.; investigation, B.C.; resources, D.C., W.B. and B.C.; data curation, D.C., W.B. and B.C.; writing—original draft preparation, D.C., W.B. and B.C.; writing—review and editing, D.C. and B.C.; visualization W.B. and B.C.; supervision, D.C. All authors have read and agreed to the published version of the manuscript.

**Funding:** The publishing procedure is funding by the JP EERA Smart Cities.

**Conflicts of Interest:** The authors declare no conflict of interest.

## References

1. The Energy Efficiency Directive (2012/27/EU) of the European Parliament and of the Council of 25 October 2012 on Energy Efficiency. Available online: http://data.europa.eu/eli/dir/2012/27/oj (accessed on 10 August 2020).
2. Wojdyga, K. Polish District Heating Systems—Development Perspectives. *J. Civ. Eng. Archit.* **2016**, *10*. [CrossRef]
3. Urbaneck, T.; Oppelt, T.; Platzer, B.; Frey, H.; Uhlig, U.; Göschel, T.; Zimmermann, D.; Rabe, D. Solar District Heating in East Germany - Transformation in a Cogeneration Dominated City. *Energy Procedia* **2015**, *70*, 587–594. [CrossRef]
4. Lygnerud, K. Challenges for business change in district heating. *Energy Sustain. Soc.* **2018**, *8*. [CrossRef]
5. Poputoaia, D.; Bouzarovski, S. Regulating district heating in Romania: Legislative challenges and energy efficiency barriers. *Energy Policy* **2010**, *38*, 3820–3829. [CrossRef]
6. Brand, M.; Svendsen, S. Renewable-based low-temperature district heating for existing buildings in various stages of refurbishment. *Energy* **2013**, *62*, 311–319. [CrossRef]
7. Connolly, D.; Lund, H.; Mathiesen, B.V.; Werner, S.; Möller, B.; Persson, U.; Boermans, T.; Trier, D.; Østergaard, P.A.; Nielsen, S. Heat Roadmap Europe: Combining district heating with heat savings to decarbonize the EU energy system. *Energy Policy* **2014**, *65*, 475–489. [CrossRef]
8. Ancona, M.A.; Branchini, L.; Di Pietra, B.; Melino, F.; Puglisi, G.; Zanghirella, F. Utilities Substations In Smart District Heating Networks. *Energy Procedia* **2015**, *81*, 597–605. [CrossRef]
9. Brand, L.; Calvén, A.; Englund, J.; Landersjö, H.; Lauenburg, P. Smart district heating networks—A simulation study of prosumers' impact on technical parameters in distribution networks. *Appl. Energy* **2014**, *129*, 39–48. [CrossRef]
10. Lichtenegger, K.; Leitner, A.; Märzinger, T.; Mair, C.; Moser, A.; Wöss, D.; Schmidl, C.; Pröll, T. Decentralized heating grid operation: A comparison of centralized and agent-based optimization. *Sustain. Energy Grids Netw.* **2020**, *21*. [CrossRef]
11. Country by Country/2015 Survey. Available online: https://www.euroheat.org/wp-content/uploads/2016/03/2015-Country-by-country-Statistics-Overview.pdf (accessed on 18 September 2020).
12. *Energetyka Cieplna w Liczbach (District Heating in Numbers)*; Energy Regulatory Office, 2016. Available online: https://www.ure.gov.pl/pl/cieplo/energetyka-cieplna-w-l/7171,2016.html (accessed on 20 September 2020). (In Polish)
13. Polskie Sieci Elektroenergetyczne. Available online: https://www.pse.pl/dane-systemowe (accessed on 20 September 2020). (In Polish).

14. Biznes Alert. Available online: https://biznesalert.pl/polska-moc-zainstalowana-mikroinstalacje-fotowoltaika-oze-energetyka/ (accessed on 16 September 2020). (In Polish).
15. Lund, H.; Werner, S.; Wiltshire, R.; Svendsen, S.; Thorsen, J.E.; Hvelplund, F. 4th Generation District Heating (4GDH). Integrating smart thermal grids into future sustainable energy systems. *Energy* **2014**, *68*, 1–11. [CrossRef]
16. Boesten, S.; Ivens, W.; Dekker, S.C.; Eijdems, H. 5th generation district heating and cooling systems as a solution for renewable urban thermal energy supply. *Adv. Geosci.* **2019**, *49*, 129–136. [CrossRef]
17. Milewski, J.; Wołowicz, M.; Bujalski, W. Seasonal Thermal Energy Storage—A Size Selection. *Appl. Mech. Mater.* **2013**, *467*, 270–276. [CrossRef]
18. Directive 2010/31/EU of the European Parliament and of the Council of 19 May 2010 on the Energy Performance of Buildings. 2010. Available online: https://eur-lex.europa.eu/eli/dir/2010/31/oj (accessed on 10 August 2020).
19. Directive 2018/844 of The European Parliament and of The Council of 30 May 2018 Amending Directive 2010/31/EU on The Energy Performance of Buildings and Directive 2012/27/EU on Energy Efficiency. 2012. Available online: http://data.europa.eu/eli/dir/2018/844/oj (accessed on 10 August 2020).
20. Rozporządzenie Ministra Infrastruktury w Sprawie Warunków Technicznych; Jakim Powinny Odpowiadać Budynki i ich Usytuowanie Warszawa. National Regulation on Technical Conditions for Buildings and Their Location with Amendments. *Dziennik Ustaw*. 13 August 2013, p. 926. Available online: https://sip.lex.pl/akty-prawne/dzu-dziennik-ustaw/warunki-techniczne-jakim-powinny-odpowiadac-budynki-i-ich-usytuowanie-16964625 (accessed on 10 August 2020). (In Polish).
21. Chwieduk, D. Impact of solar energy on the energy balance of attic rooms in high latitude countries. *Appl. Therm. Eng.* **2018**, *136*, 548–559. [CrossRef]
22. Ustawa z Dnia 20 Lutego 2015 r. o Odnawialnych Źródłach Energii z Późniejszymi Zmianami. In *Dziennik Ustaw*; 20 February 2015; p. 478. Available online: http://isap.sejm.gov.pl/isap.nsf/DocDetails.xsp?id=wdu20150000478 (accessed on 10 August 2020). (In Polish)
23. Available online: https://dane.gov.pl/dataset/797,typowe-lata-meteorologiczne-i-statystyczne-dane-klimatyczne-dla-obszaru-polski-do-obliczen-energetycznych-budynkow (accessed on 10 August 2020).
24. Rozporządzenie Ministra Infrastruktury w Sprawie Metodologii Wyznaczania Charakterystyki Energetycznej Budynku lub Części Budynku oraz Świadectw Charakterystyki Energetycznej. National Regulation on Methodology of Determination Energy Performance of Buildings or Part of Them and on Certificates on Energy Performance with Later Amendments. In *Dziennik Ustaw*; 18 March 2015; p. 376. Available online: http://isap.sejm.gov.pl/isap.nsf/download.xsp/WDU20150000376/O/D20150376.pdf (accessed on 10 August 2020). (In Polish)
25. Duffie, J.A.; Beckman, W.A. *Solar Engineering of Thermal Processes*; John Wiley & Sons: Hoboken, NJ, USA, 2013.
26. Chwieduk, D. Solar Assisted Heat Pumps. In *Comprehensive Renewable Energy*; Solar Thermal Systems: Components and Application; Kologirou, S., Ed.; Elsevier: Amsterdam, The Netherlands, 2012; Volume 3, pp. 495–528. [CrossRef]
27. Available online: https://www.meteo.waw.pl (accessed on 10 August 2020).


**Publisher's Note:** MDPI stays neutral with regard to jurisdictional claims in published maps and institutional affiliations.

*Article*

# Implementation Framework for Energy Flexibility Technologies in Alkmaar and Évora

**Nienke Maas [1,*], Vasiliki Georgiadou [1], Stephanie Roelofs [1], Rui Amaral Lopes [2], Anabela Pronto [2] and Joao Martins [2]**

1 TNO—Netherlands Organization of Applied Scientific Research, NL-2595 DA, 96800 The Hague, The Netherlands; vasiliki.geogiadou@tno.nl (V.G.); stephanie.roelofs@tno.nl (S.R.)
2 UNINOVA—Instituto Desenvolvimento de Novas Tecnologias, Faculdade De Ciências E Tecnologia, 2829-517 Caparica, Portugal; rm.lopes@fct.unl.pt (R.A.L.); amg1@fct.unl.pt (A.P.); jf.martins@fct.unl.pt (J.M.)
* Correspondence: Nienke.maas@tno.nl; Tel.: +31-646847246

Received: 30 September 2020; Accepted: 4 November 2020; Published: 6 November 2020

**Abstract:** As energy generation based on renewable resources does not always match energy consumption profiles, Positive Energy Districts (PEDs) should embody energy flexibility technologies to decrease possible negative impacts on existing grids due to, e.g., reverse power flows. As part of the EU H2020 Smart Cities and Communities project POCITYF, the cities Alkmaar (NL) and Évora (PT) aim to support the deployment and market uptake of such districts and in doing so demonstrate innovative and integrated technologies to enable flexibility in the energy system. This paper addresses implementation conditions for energy flexibility technologies that help cities to engender the expected impact and ensure replication of these technologies to other sites. It aims to guide both urban planners and technology solution providers through pitfalls and opportunities that can appear during the design and implementation of PEDs. Taking this into consideration, the RUGGEDISED innovation and implementation framework for smart city technology was taken as a starting point to describe and analyze the experiences in Alkmaar and Évora.

**Keywords:** governance; energy flexibility; positive energy districts; sustainable energy; smart city deployment

## 1. Introduction

The need for cities to become more sustainable is high and several definitions of smart cities refer to this need. A definition of a smart city is "a sustainable and efficient City with high Quality of life that aims to address Urban challenges (improve mobility, optimize use of resources, improve Health and safety, improve social development, support economic growth and participatory governance) by application of ICT in its infrastructure and services, collaboration between its key stakeholders (Citizens, Universities, Government, Industry), integration of its main domains (environment, mobility, governance, community, industry, and services), and investment in Social capital" [1]. However, retrofitting existing environments remains challenging [2]. Positive Energy Districts (PEDs) play an important role in more liveable and sustainable future cities. PEDs are districts producing energy from local and distributed renewable energy sources, presenting generation surplus over a specific balance period (typically one year) that may be transferred to areas outside a PED's boundaries [3]. One definition of PEDs by JPI Urban Europe (2019) refers explicitly to the active management of energy flows: "PEDs are energy-efficient and energy-flexible urban areas or groups of connected buildings which produce net zero greenhouse gas emissions and actively manage an annual local or regional surplus production of renewable energy" [4].

The large deployment of distributed generation based on renewable energy (RE) can increase the complexity of grid management and operation due to several factors (see, for instance, the impact of reverse power flows on distribution transformer aging [5]. Increased energy flexibility in the existing energy systems is therefore a crucial mechanism to delay costly and overdesigned adaptation of the grid infrastructure itself [6]. In this context, technology can be of great help in linking resource efficiency and flexibility in energy supply and demand with innovative, inclusive and more cost-effective services for citizens and businesses. Such technologies can integrate infrastructures like smart grids that have been piloted in several cities in FP7 and H2020 projects [7].

In Alkmaar (Netherlands) and Évora (Portugal), innovative and integrated technical solutions are being implemented in order to support the deployment and market uptake of PEDs. This research and the pilots in Alkmaar and Évora are being conducted within the EU H2020 Smart Cities and Communities project POCITYF [8]. The project started in October 2019 and includes the demonstration of solutions of a high technology readiness level (TRL $\geq$ 6) [9] for achieving flexible and efficient use of electricity in contexts with different climatic conditions and regional characteristics (technical, financial, social and legal). Alkmaar and Évora are proving grounds for innovative and integrated technical solutions for buildings and districts in which energy management systems are implemented to increase flexibility. Starting points are lessons learnt from pre-pilots at other sites, i.e., locations and buildings where individual technologies have been implemented before. The second step, which is dedicated to demonstration activities, combines technologies toward integrated systems at the building, block and district levels in the areas of Alkmaar and Évora. The third step refers to replication of these integrated systems in other selected areas of Alkmaar and Évora and in the six fellow cities: Granada (Spain), Bari (Italy), Celje (Slovenia), Újpest in Budapest (Hungary), Ioannina (Greece) and Hvidovre (Denmark).

Starting from the RUGGEDISED innovation and implementation framework developed in the EU H2020 RUGGEDISED project [10], this paper addresses implementation conditions for energy flexibility technologies in Alkmaar and Évora in order to support the achievement of expected impacts and ensure replication of these technologies within and beyond POCITYF. The technologies under consideration are ReFlex [11] for Alkmaar and flexibility control algorithms for Évora. Both aim at exploiting the energy flexibility provided by the available controllable devices (e.g., batteries or electric water heaters) at the building, block and district levels in order to achieve specific objectives (e.g., improve matching between renewable generation and energy demand or decrease peak loads).

The RUGGEDISED innovation and implementation framework is an analytical tool that helps city planners to assess important success factors in the implementation process of smart technologies well in advance. This framework focusses on smart city technologies in a broader sense without concentrating necessarily on energy flexibility technologies alone. Therefore, this paper aims to guide both urban planners and technology solution providers through pitfalls and opportunities that can appear during the design and implementation of a PED. It describes practical examples of implementation conditions seen or experienced in Alkmaar and Évora (or that were missing). This can be used to describe valuable lessons for the implementation of energy flexibility in future PEDs. This framework identifies both the implementation conditions that were of relevance and influence in either case of the two cities, and those conditions that did not play a role. In doing so, it provides insights in improving and adjusting projects and shifts the focus on conditions that matter. This paper discusses the application of the framework by presenting how it is being used in practice for the analysis of the implementation of PEDs in Alkmaar and Évora.

## 2. Materials and Methods

The RUGGEDISED framework [10] can be used to assess the implementation process of smart city projects and therefore "advises" what should be in place for successful implementation. In POCITYF, records of progress on pre-pilots and foreseen demonstration activities in Alkmaar and Évora have been taken, using questionnaires. These questionnaires focused on key technical components and specifications of the innovative solutions, on the demonstration sites (considering specific challenges related to the context), on problems and restrictions experienced and on lessons learnt (e.g., technical improvements, energy savings or socioeconomic benefits). The project partners involved with each innovative solution took care of the respective questionnaires. These project partners work for different organizations, e.g., research institutes, governments, utilities and technology providers.

Despite common challenges, the cities of Alkmaar and Évora face city- and district-specific challenges due to divergent geography, geology, demography, climate and socio-economic and cultural characteristics. These characteristics mean that urban energy transition challenges are embarked upon from different starting points and perspectives, thus enhancing the complementarity of the POCITYF solutions. Évora represents South European cities, which generally show lower investments on reducing the footprint of their households and business sector but can enjoy an abundant solar potential. Alkmaar represents West European cities that are strongly dependent on gas for electricity and heating.

In more detail, the RUGGEDISED framework provides a useful base for analyzing success factors and hindering factors for the implementation of smart technologies. In POCITYF, innovative smart city solutions are implemented in order to achieve PEDs. This framework (see Table 1) was consolidated in the EU H2020 RUGGEDISED project and allows the analysis of suppressing and enhancing factors in implementation processes of smart solutions. Such an analysis is beneficial in designing successful implementation processes, in assessing the potential project impact and in selecting aspects that need further consideration for successful implementation.

**Table 1.** RUGGEDISED innovation and implementation framework: enhancing and suppressing factors [10].

| Hardware | Software | Orgware |
|---|---|---|
| | Level of impact 1: Realization and output of smart solutions | |
| Pre-deployment assessment<br>Technology assessment<br>Impact on energy grid | Software<br>Privacy<br>Security<br>Smart Grid ICT<br>User Interfaces | Business models<br>Data and data ownership |
| | Level of impact 2: Embedded outcomes of multiple smart solutions | |
| Communicating infrastructure<br>Robustness of the system | Interoperability Dashboards | Integrated vision on the smart city<br>Smart governance<br>Windows of opportunity<br>Stakeholder management<br>Ownership<br>Business models and split incentives |
| | Level of impact 3: Upscaling and replication | |
| | | Integrated planning<br>Innovation platforms<br>Conditions for upscaling: finance, regulation (including standardization), access to information and social aspects |

The framework distinguishes three levels of impact that are needed in order to think beyond implementation of single solutions and consider the real impact of implementation, namely:

- Impact Level 1 (Realization and output of smart solutions): the first level of impact considers a single smart solution that is successfully implemented and delivers its output (it is an isolated realization of a smart solution).
- Impact Level 2 (Embedded outcomes of multiple smart solutions): A smart solution produces real output if it is well-embedded in the existing context. At the second level of impact, multiple smart solutions interact and produce outcomes because they are well-connected and efficiently work together.
- Impact Level 3 (Upscaling and replication): The third impact level is city-level outcomes. This occurs when smart solutions are taken beyond single projects and are successfully scaled-up to create smart urban structures. Real impact is made when smart solutions are replicated to other areas and projects with different contexts.

The implementation factors influencing the development of smart solutions are described per impact level. A division is made between hardware, software and orgware implementation factors. Hardware focuses on physical infrastructure topics such as energy storage, conversion and savings. Software refers to ICT, data-related factors and applications. Hardware and software must be supplemented with orgware, which refers to organizational and governance aspects such as stakeholder management, institutional and organizational arrangements and innovation platforms. Dividing impact levels and implementation factors in three categories allows for structuring the different factors that influence projects in different stages. Some factors help or hinder the realization and output of smart solutions, whilst others specifically impact the working together of multiple solutions to realize embedded outcomes. The framework shows that the importance of orgware implementation factors increases as the impact level rises. These orgware implementation factors, in turn, affect the upscaling and replication of smart solutions.

## 3. Results

One of the objectives of the POCITYF project is to deploy and validate smart energy management and storage solutions to optimize energy flows with the goal of maximizing self-consumption, reducing grid stress and valorizing flexibility services. To that end, innovative solutions to be demonstrated and replicated include several individual elements: e.g., low-temperature waste heat, innovative short- and long-term storage solutions, such as hydrogen fuel cells or electrical vehicles (EVs) coupled with stationary batteries, smart ICT solutions to interconnect the energy management system at the household, building and district levels, the virtual power plant (VPP) concept, thermal grid controllers or market-oriented building flexibility services.

Figure 1 illustrates how the individual elements corresponding to the various layers, from concrete, device-specific to abstract, domain-specific, are stacked on top of each other and interact to bring about the full implementation and impact of an integrated innovative solution. On the left side, the data and control flow is bidirectional for the full stacked solution to perform as per design, while on the right side, there is only sharing of information to gain insight into the system, evaluate different options and decide for further optimization and improvement steps. The loop is closed by feeding these insights back to the operational system in place.

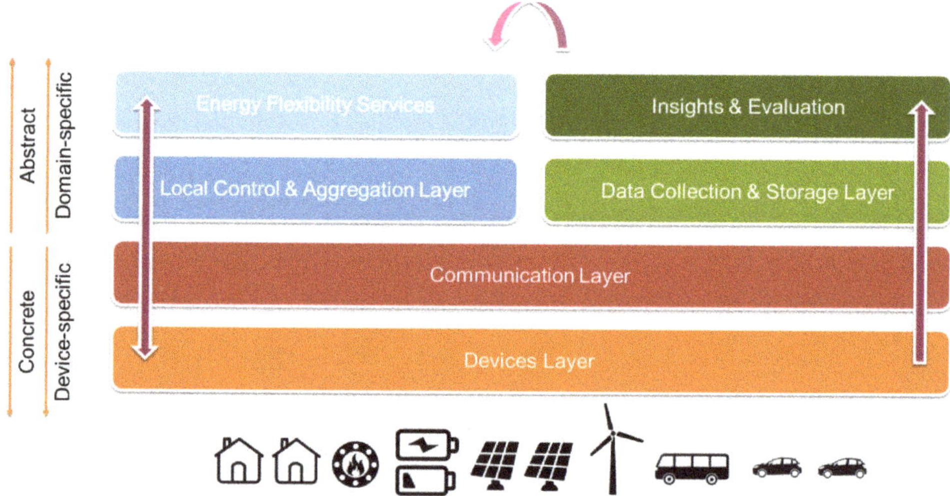


**Figure 1.** From the concrete, device-specific layer to the abstract, domain-specific one.

Taking this into consideration, Tables 2 and 3 map the information associated with the technologies under analysis (i.e., ReFlex for Alkmaar and flexibility control algorithms for Évora) based on the impact levels and implementation factors that comprise the RUGGEDISED framework described in Section 2.

**Table 2.** Mapping Alkmaar.

| **Implementation Factor** | **Explanation** |
| --- | --- |
| Level 1 realization and output | |
| **Hardware** | |
| Pre-deployment assessment | • Availability of multiple assets that may be a source of flexibility.<br>• Consideration of the objective to apply solutions that expose and valorize flexibility and potential alternatives.<br>• Feasibility study of physical implementation (all components) as modeled in silico. |
| Technology assessment | • The battery systems placed in residences are by default connected to a different phase of the 3-phase grid than both the solar panels and household energy usage. This originates from standards used in the installation sector. This means that the battery system cannot be used effectively to balance energy within the residence.<br>• Technical readiness level of components must be improved in order to use small-scale storage solutions (through VPP or vehicle-to-grid (V2G)) in a reliable way to control vital infrastructure, especially in brown field situations. Improvements in the European and international standards can ensure compatibility of the components in all situations. |
| Impact on energy grid | • Potential network constraints due to excessive flexibility operation within the system have been identified. Analysis of the implemented scenarios was carried out in which each flexibility source was individually considered; it became clear that they can be used for network congestion management up to a certain limitation. It was concluded that operation of flexibility and the different forecasting systems can also cause a congestion in the network which changes the network operation. Next to this, aggregator strategies further impact the network operation.<br>• Considering each flexibility source in relation to network limitations, especially taking into account balancing concerns and guarantee of supply.<br>• Good understanding of the layout of the low-voltage (LV) grid is a prerequisite for implementing smartification efficiently. Without these accurate data, one cannot determine where to place the measurement equipment in the grid. State estimation is needed in order to be able to draw conclusions about the current situation of the grid.<br>• VPP in trade mode at times might have a negative effect on grid balance given the current pricing mechanism. There are instances where energy is sold, and thus delivered to the grid, during moments of local energy generation. This can increase the peak load on the low-voltage grid in the districts.<br>• By using the VPP for energy trading, based on forecasts of energy consumption, PV generation, the energy prices on the trading market and the availability of the battery, overall energy costs can be decreased. A low density of VPP contributors in the LV grid (less than 5% of residences were equipped with battery systems) showed little impact on the grid. The density of the VPP contributors should be increased to notice the effect on the grid and to have accurate predictability of average household behavior.<br>• The distribution system operator (DSO) can alleviate peak loads on the grid by using small storage (battery) systems. These batteries need to be operated in an inherently grid-supporting mode, and peak load shaving requires fast and reliable control mechanisms. From a business perspective, adequate financial stimuli are needed. The DSO should have a mandate to overrule the settings of the VPP to ensure grid stability and prevent a power outage. |

**Table 2.** *Cont.*

| Implementation Factor | Explanation |
| --- | --- |
| Level 1 realization and output | |
| **Software** | |
| Privacy | • Including privacy by design in order to meet both national and European standards and adhere to GDPR guidelines. In doing so, privacy concerns will not pose obstacles to implementation. |
| Security | • Taking cyber security issues into account as they may arise. Including privacy and security by design so that they will not become an obstacle for implementation and the implementation will meet the required national and European regulations and standards. It can be helpful to follow the eight principles of privacy by design [12]. |
| Smart grid ICT | • Internet of things software from ICT.eu links all electricity consumers per platform. The data are made accessible via the cloud, giving the energy supplier remote control. Via this platform, everything can be connected per household that uses or generates energy. |
| User interfaces | • Using future-proof standards (even with low maturity) to implement new use cases and future business models. |
| **Orgware** | |
| Business models | • Cost reduction of grid operations as a result of unlocking and exploiting flexibility.<br>• New revenue streams potential from valorizing flexibility.<br>• Analysis of pricing mechanisms and market liquidity has been carried out as well as a usability analysis.<br>• Bundling multiple smaller devices to increase accessibility to the trade market. |
| Data and data ownership | • Handled according to the principle "as open as possible, as closed as necessary". |
| Level 2 embedded outcomes of multiple smart solutions | |
| **Hardware** | |
| Communicating infrastructure | • Continuous internet connection in order to communicate in real time between assets, the network and the energy market. |
| Robustness of the system | • Fast response to disconnection.<br>• Backup facility.<br>• Risk management: what goes wrong if the system is down or unavailable? Taking measures appropriate to the risks (system overload or just some missing data).<br>• The placement of bidirectional charging stations is currently not a standard job. They are still far more expensive than standard charging stations, they need a reliable internet connection to function and during the City-zen project, problems with the hardware were encountered frequently.<br>• The reliability of OT/IT connections through 4G and Wi-Fi communication networks used for VPP and V2G applications is lower than that of the grid itself. The reliability needs to increase to match the reliability of the grid before a VPP or V2G can be used as part of the vital infrastructure balancing loads on the LV grid. |

**Table 2.** *Cont.*

| Implementation Factor | Explanation |
|---|---|
| **Software** | |
| Interoperability | • ReFlex is compatible with several open smart grid standards such as EFI/S2, USEF and OCPI and is easy to implement in an aggregator's existing architecture. Still, bidirectional translation of asset protocols to EFI is a necessary configuration step to enable reliable communication.<br>• Reliance on standards instead of proprietary protocols to allow interoperability is paramount.<br>• The maturity of standards for interoperability is low; as such, in every single project, protocols and interfaces need to be (re)determined between different suppliers of technology. |
| Dashboards | • Management and administration dashboards are included in ReFlex. |
| **Orgware** | |
| Integrated vision on the smart city | • A local needs assessment showed that good accessibility with a balanced mobility system and a healthy, clean, safe, economic future-resilient, green and climate-adaptive city where everyone can happily live, work and recreate were important.<br>• Flexibility solutions can help the DSO in operating the network and prevent investments in grid infrastructure. |
| Smart governance | • Insights gained regarding both the flexibility potential and its value within the selected pilot areas can further support decision making and planning. |
| Windows of opportunity | • Implementation of ReFlex instead of investments in grid reinforcement: awareness of DSO that avoiding grid reinforcements is in their own interests and that ReFlex is valuable for network operation so that different problems and interests of stakeholders are aligned.<br>• The higher the number of flexible assets, including EVs, the more value can be extracted through ReFlex deployment.<br>• Local communities can become sustainable by being as self-proficient as possible while monitoring energy use at the individual and district levels and using this information for behavior change. The business district of Boekelemeer could initiate a Citizen Energy Community (CEC) or a Renewable Energy Community (REC). According to the EU directive [13], this could improve their position in the local energy market. They can use this institutional framework to represent interest of the local community in the park management. |
| Stakeholder management | • A guiding role for the municipality in building the relationship between stakeholders.<br>• Involvement of and willingness to engage from residents is crucial in relation to user acceptance. Support of homeowners is needed but they lack information and knowledge on energy flexibility as well as on practical consequences and potential benefits.<br>• Different reward mechanisms for VPP participants should be tested to determine the financial and general economic value of these systems. |
| Ownership | • Multiple aggregators in order to guarantee freedom of choice for citizens (selling energy flexibility) and the DSO (congestion management).<br>• Citizen (local) communities, municipality and housing corporations can have ownership of the applications and should be determined per case. |
| Business models and split incentives | • Analysis of market liquidity and pricing mechanisms. |

**Table 2.** *Cont.*

| Implementation Factor | Explanation |
| --- | --- |
| Level 3 upscaling and replication | |
| **Orgware** | |
| Integrated planning | • It is important that processes continue to be managed and it is essential that the involved key figures remain available at crucial positions.<br>• There was no technical supplier as a consortium partner or as official project partner. It is recommended to seek active collaboration with a supplier of technical components. |
| Innovation platforms | Important element to create a successful local innovation ecosystem is a problem-solving environment providing structure and practices of innovation management, training and tools as well as financial resources for innovation. In Alkmaar, this has been initiated but from an external perspective not well anchored in the city organization. |
| Conditions for upscaling: finance, regulation (including standardization), access to information and social aspects | Market development:<br>• Aggregator strategies can potentially greatly impact upon the energy system since their objective of pursuing economic maximization of their assets may not always align with optimization of operations for said assets; for example, stationery batteries or EV operation may be masked by the aggregator whose portfolio includes such assets by revealing only the worst case of the day instead of their real-time state. Nonetheless, it also opens the door for other aggregators to dive in with their flexibilities. This might result in a game theory problem where in reality the aggregators would self-regulate themselves. Since there is no flexibility market, the single aggregator in one network can act as a monopolist.<br>• Certain capabilities of both the VPP and V2G system cannot be utilized due to regulatory barriers, sector standards or both. *See also below.*<br>Regulation and standardization:<br>• Similarly, regulation must be adapted to enable better use of the current and future electricity system possibilities such as use of flexibility and congestion management. Due to current regulations and sector standards, certain capabilities of both the VPP and V2G system cannot be utilized. Examples are the use of the V2G system for emergency power and the self-consumption of energy using the VPP by connecting the battery and solar panels of a household to the same phase.<br>• Implementation of a variable tariff system in the NL in which energy taxes are coupled to the energy price and network tariff. Changing the Dutch taxation tariff system can pave the way for using (local) flexibility and prevent more grid reinvestments than necessary. This can be accomplished with, e.g., variable network tariffs. Coupling the energy tax not to the energy amount but the price of energy and network tariffs can have positive effects. Further research is needed in order to determine which tariff system is best applicable.<br>Social aspects:<br>• Finding and retaining participants for VPP and V2G projects is a time-consuming task. Active participation and a substantial time investment are required which is not always possible or desired. In addition, participants sometimes move or retract from the project due to life events. The promoted definitions of CEC and REC from the EU Green Energy Package can provide a much-needed boost on this area. |

**Table 3.** Mapping Évora.

| **Implementation Factor** | **Explanation** |
|---|---|
| Level 1 realization and output | |
| **Hardware** | |
| Pre-deployment assessment | • Definition of objectives for the flexibility control algorithms.<br>• Characterization of controllable devices providing energy flexibility.<br>• Preliminary tests using software tools. |
| Technology assessment | • Flexibility control algorithms were developed based on knowledge collected during the participation on the working group "Annex 67" of the Energy in Buildings and Communities (EBC) program from the International Energy Agency (IEA).<br>• The deployment of flexibility control algorithms has been considered to (i) improve PV self-consumption and buildings' self- sufficiency; (ii) explore the existing energy flexibility available in water heaters to decrease electricity costs; or (iii) characterize and use of the energy flexibility provided by water pumping and storage systems to reduce electricity costs and support power systems. |
| Impact on energy grid | • The impact related to the deployment of flexibility control algorithms on existing grids should be considered before the respective real-world implementations.<br>• This impact assessment should consider different scenarios (e.g., seasons of the year) and can be conducted using simulation tools and specific information about the grids under consideration.<br>• The acquisition of information about the grids under consideration can be challenging if the operator is not included in the process. |
| **Software** | |
| Privacy | • Data privacy issues in relation to the framework for data exchange and related roles and responsibilities were not considered in the pre-pilots of flexibility control algorithms, however technical issues supporting the exchange of data in a secure and interoperable manner are easy to integrate, respecting Portuguese law. |
| Security | • Data security in relation to the framework for data exchange and related roles and responsibilities were not considered in the pre-pilots, however technical issues supporting the exchange of data in a secure and interoperable manner are easy to integrate. |
| Smart grid ICT | • All flexibility-related data are available on the cloud and can be shared with authorized third parties. |
| User interfaces | • User interfaces were developed for the pre-pilot installations and they can be easily integrated with high-level SCADA (or similar) solutions. |
| **Orgware** | |
| Business models | • Depending on the objectives of the considered DSM measures, the flexibility control algorithms have resulted in PV self-consumption improvements up to 30% or in peak load reduction up to 10%.<br>• Flexibility control algorithms have also been used to take advantage of time-of-use tariffs in order to shift energy consumption to less expensive periods, resulting in savings up to 40%.<br>• New revenue schemes based on flexibility usage provide grid operation cost reduction. |
| Data and data ownership | • In the pre-pilot, data are private but available for authorized third parties during the assessment phase. |

**Table 3.** *Cont.*

| Implementation Factor | Explanation |
| --- | --- |
| Level 2 embedded outcomes of multiple smart solutions | |
| **Hardware** | |
| Communicating infrastructure | • As a software solution, flexibility control algorithms will not have a physical presence in the Positive Energy Blocks (PEBs). This solution will run in servers located at partners' facilities, communicating with the devices installed at the demonstration sites. |
| Robustness of the system | • Servers will have a backup, in case of hardware failure.<br>• On-site devices will have a distributed software solution in order to increase the robustness of the overall solution.<br>• Backup procedures will be considered. |
| **Software** | |
| Interoperability | • Besides the building level control of the available energy flexibility, flexibility control algorithms can be used at district level to support the coordinated operation of other technologies providing energy flexibility to achieve building-level objectives while targeting improvements at the district level (e.g., peak load reduction).<br>• The deployment of the flexibility control algorithms requires the interaction with existing controllable devices to perform the required monitoring and control activities. In POCITYF, this interaction will be supported by APIs made available by the controllable device itself (e.g., direct interaction with energy routers [14] or by other intermediate systems (e.g., indirect interaction with 2nd life batteries through energy management systems).<br>• Usage of standard protocols. |
| Dashboards | • Dashboards will be provided by outside solutions that offer a graphical interface for the users. Foreseen devices include: energy routers, 2nd life batteries, electrical water heaters or freezing storage systems. |
| **Orgware** | |
| Integrated vision on the smart city | • An assessment at district level should be carried out to define higher-level objectives to be considered by the flexibility control algorithms in order to support the achievement of a specific smart city vision. |
| Smart governance | • Smart governance would increase the usage of flexibility control algorithms at the municipality scale (and citizen engagement would be enlarged) in order to increase sustainability and climate change mitigation. |
| Windows of opportunity | • Deployment of flexibility control algorithms (considering the energy flexibility provided by, e.g., existing residential batteries and other controllable devices such as electric water heaters) instead of investments in grid reinforcement. |
| Stakeholder management | • Local communities' engagement will be a key element in the deployment and proper operation of flexibility control algorithms. |
| Ownership | • Residents and municipality will have access to the systems on their premises. |
| Business models and split incentives | • Specific business models and incentives should be developed to accommodate potential conflicts between building- and district-level objectives. |

**Table 3.** *Cont.*

| Implementation Factor | Explanation |
| --- | --- |
| Level 3 upscaling and replication | |
| **Orgware** | |
| Integrated planning | • High-level analysis will be needed in order to choose the best locations for upscaling the installation of controllable devices, combined with global flexibility control algorithms. |
| Innovation platforms | • Demonstration partners will have preferential access to all developments/features of the innovative elements and the opportunity to shape them according to their needs. |
| Conditions for upscaling: finance, regulation (including standardization), access to information and social aspects | • Mobilization of (and networking with) key stakeholders to create strong links and foster engagement of the different target groups at the local, national and international levels, which also represent the enablers for further upscaling and replication of the POCITYF solutions.<br>• Adapt regulations and market conditions in order to fully use the potential of flexibility, providing market-oriented building flexibility services.<br>• Find attractive ways of engaging citizens. |

## 4. Analysis

The RUGGEDISED framework allows for conducting an analysis of important implementation conditions in the cities of Alkmaar and Évora. This section presents the respective main observations.

- Energy flexibility management relies on both energy flexibility characterization and its respective use. This means that the energy flexibility system is based on two main steps. First, it is necessary to get insights (mainly related to flexibility characterization) on the flexibility usage potential and application. Secondly, this flexibility potential will be applied while managing and controlling the energy system.
- Complex energy services like peer-to-peer (P2P) energy trading and VPP, management aggregation services and an energy flexibility marketplace are built upon different individual components. These lower layer(s) components must be functioning and deliver their output in order to provide, possibly stacked, smart (energy) services. As such, Level 1 functioning of such individual components is a prerequisite for smart energy services to exist and be deployed.
- P2P platforms are deemed important from the abstract scenario perspective. However, development mostly starts from a technical solution perspective, while end-user and other stakeholder requirements are paramount in development. Citizen perspectives are often overlooked. Citizens mostly experienced P2P as a burden or a complex, technical solution. According to Alkmaar's experience, relevant stakeholders were not interested, and they lacked knowledge and information, while developers assumed citizens were eager for these technologies. Therefore, engaging all stakeholders and especially end-users already from the conceptualization phase of energy flexibility systems will benefit end-user satisfaction. To achieve this, it is important to bring developers and end-users together so that requirements and needs of the local community are considered and in the long term to ensure widespread and sustainable uptake. Most devices cannot provide flexibility without the involvement of users. Therefore, they are crucial in generating flexibility in energy systems. It is undeniable that the perspective of end-users is important in every phase from conceptualization to development, deployment and operations.
- Transition from Level 1 to Level 2 is normally funded by subsidies. However, subsidized projects in which energy flexibility technologies are piloted have different timeframes that do not always match with the daily business of industrial and business partners. Next to this, the research or development objectives of subsidized energy projects do not per se align with business goals

(in time, budget and purpose). This results in the delivery of flexibility technologies that function (Level 1) but will not necessarily graduate to achieve Level 2 and Level 3 impact; as such, upscaling and replication are not induced. Single flexibility technologies are delivered but this does not result in energy flexibility as a whole or flexibility in smart city systems. At the community or neighbourhood level, subsidies are effective in reaching goals, but in order to achieve large-scale impact (Level 2 and Level 3), combined strategies and widespread collaboration are needed.

- Energy flexibility usage at higher levels should take into consideration the control objectives of previous levels. This is the case when the energy flexibility of a specific device is being used at Level 1 to, e.g., reduce electricity costs and then it is used to, e.g., reduce electricity consumption $CO_2$-related emissions at Level 2. Therefore, the considered business models should properly address the transition to upper limits in terms of objectives of energy flexibility use. Following the previous comment, one can infer that each level may have different objectives which may not only be disconnected from each other but also in some cases directly conflicting. Business models need to be designed in order to address the different strategies that can be applied to the same individual devices at different levels.
- User engagement is key in exploiting the energy flexibility provided by several types of controllable devices as the comfort needs should always be respected. Additionally, the characterization of the energy flexibility provided by some devices (e.g., white appliances) might require the direct user interaction to define the respective operation boundaries. This constitutes a constraint to be considered at all levels.
- It is of upmost importance to regard investments in flexibility usage as win–win situations. For effective upscaling to take place, all involved stakeholders (e.g., DSO, municipalities, end-users) must have a clear idea about the advantages of investing in flexibility.
- There is no possibility to have accurate modeling without data, and data cannot be used if privacy and security issues are not considered. Agile data management should be considered in order to take the best profit from the considered technologies. A good example is the smart metering data, where often end-users do not take advantage of their full potential.

## 5. Conclusions

In this paper, the RUGGEDISED framework was applied as the starting point to evaluate the implementation conditions of energy flexibility solutions within the POCITYF project. Indeed, the selected framework can be leveraged to provide a first analysis of both the success and hindering factors for the implementation of smart technologies while taking a holistic approach that includes not only technological perspectives but also organizational and systemic ones.

The main recommendations and findings for implementing energy flexibility-related technologies at the PED level are summarized below.

- Advanced innovative services and new business models, such as P2P trading and VPP, are quite complex technologies and procedures. They can only be implemented at Level 1 if the individual necessary components such as storage devices (for example, batteries) have passed Level 1.
- Services like P2P can only be introduced when user needs have been considered and user requirements are met. Therefore, involving end-users in the development of the services is key to reach goals and ensure continuous uptake from a bigger group of clients. It is the developer's responsibility to consult end-users and the entire stakeholder value chain.
- The two main functions of the energy flexibility system must be connected. Data insights will have to feed back into the management and control process. The results of the control process will have to be monitored and provide insight on the impact of the introduced control mechanism. Without this feedback loop, the installed energy flexibility system will not be able to deliver the potential optimal flexibility. Ergo, if there is impact, it is by coincidence and therefore it cannot be sustained in the long term. Feedback can change the characterization and the use. A feedback

loop enables deeper understanding of the sources of error, that can either be found in the control (use) or in the characterization function. For acceptance of energy flexibility, it is important to close this loop. Insights delivered via tracking key performance indicators (KPIs) are needed both on rational arguments (economic and energy metrics) and on perception arguments (e.g., satisfaction of citizens).

- A legal framework will have to be updated to realize the full-scale potential of flexibility, because the current legal framework does not allow for using and implementing flexibility. We may apply it as a local, pilot-based tool. However, if one wishes to exploit its potential in full, a consistent and encompassing legal framework is necessary.
- In order to implement energy flexibility systems in the built environment and to eventually scale up at district and city levels, gaps between daily practice and subsidized projects should be closed. Goals of all partners must coincide with the aims of pilot projects. This is crucial for realizing Level 3 outcomes but also for Level 2.
- The absence of an adequate legal framework is a very important issue and not only for specific energy flexibility technologies. Systemic hiccups introduced by a lack of appropriate or mature legal instruments should be overcome beyond pilot project boundaries which in any case are funded by subsidies whose power to influence directly relevant legislation is limited (both in scope and time).

The experiences of Alkmaar and Évora result in an easy-to-use methodology for cities to deal with energy flexibility technologies and create optimal conditions for their successful implementation and integration. In the coming period, demonstration of the related energy flexibility technologies is being carried out in both cities at the block and district levels. Starting from the analysis, progress is to be tracked especially focused on achieving Level 1 and moving forward to higher impact levels. This effort will also support the replication process not only in other areas within Alkmaar and Évora but also the so-called "fellow" cities participating in the POCITYF project.

## 6. Practitioners Review by Roel Massink MSc

The authors of this paper asked me to provide my view on the methods and results of this research paper in my capacity as Project Coordinator for the H2020 Smart City Lighthouse project IRIS and innovation manager for the Municipality of Utrecht. I am happy to fulfil this task and I thank the authors for the carefully presented research results based on actual demonstration. This review is based on the experiences collected in the IRIS Smart Cities project, a similar project to RUGGEDISED and POCITYF. In IRIS, an integrated project approach is demonstrated: in the Kanaleneiland district (Utrecht, The Netherlands), near zero-energy-efficient building retrofitting is connected to the development of a smart energy and mobility system integrating PV panels with stationary and V2G storage. The smart solutions in IRIS Smart Cities Utrecht mostly all fall within Level 2 of the RUGGEDISED implementation framework. Below, some general reflections based on these experiences toward the RUGGEDISED implementation framework and the conclusions of the paper can be found.

The structure of the RUGGEDISED framework is very much in line with what is happening in practice in the IRIS Smart Cities project and further development of the framework in practice is appreciated. Within the IRIS Smart Cities project, the smart solutions are implemented along transition tracks and structured through a maturity assessment in the following categories: (1) pre-pilots: solutions that have been tested on a small scale in a pilot project; (2) integrated solutions: multiple smart solutions that are demonstrated as integrated solutions at a larger scale at the demonstration site of the project; and (3) replicated solutions: through either the copying of a successful integrated solution to other sites or by upscaling of the integrated solution in the same city or region. This categorization provides the IRIS partners and stakeholders a framework and a roadmap for all smart solutions to be developed, demonstrated and scaled-up. The RUGGEDISED framework is valuable and will offer practitioners at public authorities, grid operators, solution providers and knowledge institutes a coherent overview of what is important for implementation. However, below I have some suggestions in view of the practical applicability of the RUGGEDISED framework. First, smart solutions are integrated solutions by nature. To ensure that the smart solutions (that are realized in isolation; Impact Level 1) meet the requirements of an embedded outcome of multiple smart solutions (Impact Level 2), it is suggested that an integrated approach with all implementation factors is taken into account. To ensure such an integrated (and systemic) assessment throughout the development of smart solutions, an iterative execution of the implementation framework could be proposed. This means that at Level 1, the implementation factors of Level 2 and Level 3 are also considered. Assessment of implementation factors in Level 2 and Level 3 includes more detailed and specific information once the development of smart solutions progresses.

The following three reflections are supportive of including higher-level implementation factors at the start of the pilot projects.

The paper highlights the importance of aligning partner goals with pilot project aims. This should be seen as a conditional factor for achieving progress at the different impact levels. An example from IRIS Smart Cities Utrecht is the development of a bidirectional charging ecosystem for grid flexibility and mobility services. Currently, this integrated solution consists of multiple smart solutions (bidirectional charging infrastructure, bidirectional enabled electric vehicles, stationary battery storage, energy management system) that are being connected to provide embedded outcomes (Level 2). The current state of development is a result of careful stakeholder management from the start of the pilot projects (already pre-IRIS). Companies, knowledge institutes, grid operators and public authorities aligned their roadmaps and activities in pilot projects under the lead of an ambitious SME (Level 1) and now this is leading to embedded outcomes in the IRIS demonstration project (Level 2). Furthermore, scaled adoption in public procurement documents of electric charging infrastructure now paves the way for city-wide flexibility services of electric vehicles (Level 3). An important success factor was/is stakeholder alignment from the start of the pilot projects. Based on this experience, it is argued that stakeholder management (or rather stakeholder alignment) should take position already at Level 1 to ensure that partner roadmaps are aligned to move the smart solution to Level 2 and 3.

The paper also points towards the requirement of end-user involvement (or co-creation) in the development of smart solutions and could argue for the inclusion of a new implementation factor in the RUGGEDISED framework. This could be "end-user satisfaction" or similar. This is a valid point; often smart solutions are hampered because end-user needs were not considered well enough in the original approach. A design-thinking (or another systemic) approach could offer smart solution developers tools to better involve end-users. A practical notion that requires attention here is the use of grant subsidies in this innovation framework (as the authors also refer to in moving from Level 1 to Level 2). It is recommended that subsidy programs, grant applicants and consortia put more attention into making end-user involvement more explicit in call texts and subsequently also allow more flexibility in the implementation of the grant project based on changes offered by end-users.

Next to this, the paper explains the requirement of a legal framework that supports the transition from pilot projects to scaled solutions. This is true as well for the IRIS Smart Cities project. Especially within the field of smart energy projects or services like peer-to-peer, a supportive legal framework is needed for scaled adoption. The exploitation of grid flexibility services is hampered by legal constraints. Research from the IRIS project shows that bidirectional charging services are discouraged because of double energy taxation (for each charging and discharging cycle, energy tax needs to be paid on either the stored or consumed kWh). This significantly hampers the realization of profitable business models and commercial scaling of these integrated smart solutions. Therefore, it could be argued that regulation/the legal framework is already introduced as an implementation factor in Level 2 of the framework to ensure that, in time, legal constraints are targeted in a concerted action by smart cities.

Finally, what the paper does not explain but what is highly valuable for practitioners is guidance on how the implementation framework leads into an implementation process. The presented implementation framework provides an assessment of implementation factors related to smart solutions but does not directly translate the assessment into implementation guidelines. Further guidance on the implementation pathway resulting from the assessment could support practitioners in applying this framework more easily.

**Author Contributions:** Conceptualization, N.M., V.G. and S.R.; methodology, N.M. and S.R.; validation, N.M., R.A.L. and J.M.; formal analysis, N.M., V.G., S.R., R.A.L. and J.M.; investigation, N.M., V.G., S.R., R.A.L., A.P. and J.M.; resources, S.R. and R.A.L.; writing—original draft preparation, N.M., V.G., S.R., R.A.L., A.P. and J.M.; writing—review and editing, N.M., V.G., S.R., R.A.L. and J.M.; visualization, V.G.; supervision, N.M. All authors have read and agreed to the published version of the manuscript.

**Funding:** This research was funded by POCITYF, (FP7 grant agreement N° 864400). And The APC was funded by EERA Joint Program Smart Cities.

**Acknowledgments:** This paper is based on information collected during the initial phase of EU H2020 Smart Cities and Communities project POCITYF (grant agreement N° 864400). The authors would like to express their gratitude to Roel Massink MSc, coordinator of the H2020 Smart City Lighthouse project IRIS, for reviewing the methods and results of this research and for all the suggestions provided.

**Conflicts of Interest:** The authors declare no conflict of interest.

## References

1. Mosannenzadeh, F.; Vettorato, D. Defining smart city. A conceptual framework based on keyword analysis. *TeMA-J. Land Use Mobil. Environ.* **2014**. [CrossRef]
2. Dixon, T.; Eames, M. Scaling up: The challenges of urban retrofit. *Build. Res. Inf.* **2013**, *41*, 499–503. [CrossRef]
3. Shnapp, S.; Paci, D.; Bertoldi, P. *Enabling Positive Energy Districts across Europe: Energy Efficiency Couples Renewable Energy*; EUR 30325 EN; Publications Office of the European Union: Luxembourg, 2020; ISBN 978-92-76-21043-6. [CrossRef]
4. Positive Energy Districts (PED). JPI Urban Europe. Available online: https://jpi-urbaneurope.eu/ped/ (accessed on 25 October 2020).
5. Lopes, R.A.; Magalhães, P.; Gouveia, J.P.; Aelenei, D.; Lima, C.; Martins, J. A case study on the impact of nearly Zero-Energy Buildings on distribution transformer aging. *Energy* **2018**, *157*, 669–678. [CrossRef]
6. IEA. World Energy Outlook 2019. IEA, Paris. Available online: https://www.iea.org/reports/world-energy-outlook-2019 (accessed on 25 October 2020).
7. Available online: https://www.h2020-bridge.eu/participant-projects/ (accessed on 25 October 2020).
8. H2020 POCITYF Project. Available online: https://pocityf.eu (accessed on 2 September 2020).
9. Technology Readiness Levels (TRL). HORIZON 2020—WORK PROGRAMME 2014–2015 General Annexes. Available online: https://ec.europa.eu/research/participants/data/ref/h2020/wp/2014_2015/annexes/h2020-wp1415-annex-g-trl_en.pdf (accessed on 2 September 2020).
10. Slob, A.; Woestenburg, A. (Eds.) *Overarching Innovation and Implementation Framework*; D1.2 Ruggedised Smart City Lighthouse Project; TNO: The Hague, The Netherlands, 2017; pp. 10–13.
11. ReFlex Technology for Smart Energy Services. TNO. Retr. Available online: https://www.tno.nl/en/focus-areas/techtransfer/licenses/reflex-technology-for-smart-energy-services/ (accessed on 28 October 2020).
12. OECD Guidelines on the Protection of Privacy and Transborder Flows of Personal Data. Available online: https://www.oecd.org/internet/ieconomy/oecdguidelinesontheprotectionofprivacyandtransborderflowsofpersonaldata.htm (accessed on 2 September 2020).
13. 2020 Climate & Energy Package. Climate Action—European Commission. Available online: https://ec.europa.eu/clima/policies/strategies/2020 (accessed on 2 September 2020).
14. Roncero-Clemente, C.; Vilhena, N.; Delgado-Gomes, V.; Romero-Cadaval, E.; Martins, J.F. Control and operation of a three-phase local energy router for prosumers in a smart community. *IET Renewa. Power Gener.* **2020**, *14*, 560–570. [CrossRef]

*Article*

# Analysis of the Development and Parameters of a Public Transport System Which Uses Low-Carbon Energy: The Evidence from Poland

**Justyna Patalas-Maliszewska * and Hanna Łosyk**

Institute of Mechanical Engineering, University of Zielona Góra, 65-417 Zielona Góra, Poland; losyk.hanna@gmail.com

* Correspondence: j.patalas@iizp.uz.zgora.pl

Received: 29 September 2020; Accepted: 2 November 2020; Published: 4 November 2020

**Abstract:** Efforts toward a low-emission economy constitute a common challenge for Polish cities. Solutions are being sought to support Polish, medium-sized cities, that is, cities with about 140,000 inhabitants, to implement and develop low-carbon energy in their public transport systems. This paper proposes and explores a sustainable urban development card for a Polish city, namely, Zielona Góra, the use of which will enable the effects of a public transport system using low-carbon energy to be monitored. This research was based on the two main areas of analysis of a system of low-carbon energy and public transport and were formulated as: (1) Sustainable Development Goals (SDGs) and (2) Indicators of the Satisfaction Rate of Public Transport Passengers (SPTP). This paper used literature studies to determine SDGs as well a questionnaire-cum-survey, which was conducted on a sample of 1022 public transport passengers in Zielona Góra, Poland, to determine SPTP. The results were verified by a real case study of a Polish city, which, in 2019, had the largest fleet of electric buses in Poland; a statistical analysis was also conducted using correlation coefficients. It was determined that the proposed approach allows for low carbon energy public transport to be constantly monitored and analyzed. In the long run, this could be a good benchmark as to how cities might improve their level of sustainability.

**Keywords:** electric buses; greenhouse gas emissions; low carbon energy public transport; public transport; sustainable development; Polish city

---

## 1. Introduction

Transport is the main source of greenhouse gas emissions. According to data provided by the International Energy Agency (IEA), activities should be implemented in the transport sector in the coming decades, in order to reduce $CO_2$ emissions by 70% in developing countries and by 80% in developed countries in advance of 2050 [1].

### *1.1. Smart Energy for Communities*

In response to the challenges of the modern world such as the energy crisis, rapid climate change, and environmental issues related to the use of fossil fuels and their volatile prices, the undertaking of initiatives in line with the concept: "community energy" is essential for the development of renewable energy sources and local energy management [2–4].

The idea of "community energy" is often presented as activities where communities exhibit a high degree of control of the energy project as well as collectively benefitting from the outcomes either of energy-saving or revenue-generation [5]. Another comprehensive definition is also that presented by Klein and Coffey [6], which presents community energy as "a project or program initiated by a group of people, united by a common, local geographic location such as a town- or smaller- and/or a set of common interests, in which some, or all, of the benefits and costs of the initiative are applied to this same group of people and which incorporate a distributed energy generation technology for electricity, heat, and transportation, based on renewable energy resources such as solar, wind, water, biomass, geothermal and/or energy conservation/efficiency methods/technologies". An analysis of the literature has shown that the main characteristics of community energy include [7] social initiatives, which are largely driven by social need, motivation, and values. The values adopted vary and may also vary in nature; a sense of a high degree of community results in a high degree of commitment to the community in management and decision-making. Initiatives should aim at a fair distribution of values, costs, and risks; society is not only focused on maximizing the economic benefits.

In the field of smart energy for communities supporting the transition to low carbon energy, an important concept is the virtual power plant (VPP), which appeared as early as the late 1990s. One of the commonly cited definitions of VPPs is that of Pudjianto and others [8], where a VPP is considered to be "a flexible representation of a portfolio of distributed energy resources (DER), which aggregates the capacity of many diverse DER to create a single operating profile from a composite of the parameters characterizing each DER and incorporating spatial (i.e., network) constraints". Literature on the subject distinguishes two classifications of VPPs: commercial and technical due to their roles in the energy system [8]. Among the most popular commercial solutions, smart energy communities supporting a transition to low carbon energy stand out: solar panels, solar farms, and heat pumps.

One example of community energy is the large scale introduction, in many city centers, of electric buses that run on batteries. The biggest advantage of battery-run, electric buses, is zero local emissions as well as the possibility of reducing their operating costs [9]. However, compared to conventional vehicles powered by internal combustion engines, electric buses are somewhat limited when it comes to storing energy since this affects the driving range. When analyzing the driving range, attention should not only be paid to the capacity of the battery used, but also to the environmental conditions and to the conditions of the vehicle including the use of auxiliary, vehicular devices, and the weight of the vehicle as well as the road conditions [10]. The literature on the subject, in connection with uncertainty, the range of electric vehicles running on batteries, presents methods for predicting energy consumption including the results of experiments allowing such as the range of the battery used to be accurately estimated when adopting various parameters of friction, the weather conditions, and/or operation [11,12], and the data-driven methods such as the use of neural networks to determine the energy consumption of electric vehicles [13]. A significant number of models presented take into account dependent variables, which are particularly important when considering city buses. Next to the driving range, attention should be paid to the problem of the growing demand for materials to build electric batteries [14]. As pointed out in [15], the current production of electric vehicles is not close to the total production of passenger cars. In order to increase production, the scenario of increasing the extraction of certain metals, in particular lithium, but also dysprosium, terbium, platinum, neodymium, tantalum, and palladium [15] should be adopted. Due to the popularity of the use of lithium batteries, this material is considered potentially problematic, not only due to its limited resources, but also because it is time-consuming and expensive to recycle; recycling is not fully developed at present. However, it should be emphasized that there are different types of batteries available on the market, which clearly implies risks and limitations. Our research focused on analyzing the effectiveness of implementing electric buses in cities in the context of improving sustainability levels.

Another important parameter, when comparing electric buses to traditional buses with internal combustion engines, is noise. Today, transport is the main source of noise in urban areas, which negatively affects both the health and the general emotional state of residents. Electric buses practically do not emit sounds at low speeds, which significantly affects the quality of life of residents. However, this parameter is also considered in terms of road safety, since it is the noise emitted by moving vehicles that is a key information factor for road users [10]. There has been a higher proportion of road accidents involving pedestrians and cyclists caused by collisions with electric/hybrid vehicles [14]. Woodcock et al. [15] tested electric buses in UK cities for noise emissions. In the test results, drivers of electric buses agreed that the dearth of noise and vibration in the cab improved both their comfort and that of the passengers. At the same time, they pointed out that the main problem of noise emission is the safety of pedestrians on the roads since safety, as such, is mainly based on hearing.

Our research, as pointed above, focused on analyzing the effectiveness of implementing electric buses in the city. Therefore, a literature analysis on measuring the effectiveness of the use of electric buses in cities is the measurement of greenhouse gas emissions and energy consumption. The results presented by [16] indicate greenhouse gas emissions for transit buses at 211 g $CO_{2eq/km}$, while the implementation of electric buses aims for the maximum reduction of emissions of $CO_{2eq/km}$ when running. Liu Q. and others [17] presented research on measuring the emission flow of buses powered by different fuels when accelerating from a bus stop in Gothenburg, Sweden. The analysis showed that electric buses emitted significantly fewer harmful gases when starting from a bus stop when compared to diesel buses; this, in turn, affects the amount of PM dust produced and subsequently inhaled by passengers and residents. The undoubted benefit of the implementation of electric buses in affecting the quality of life of residents is the reduction of noise in cities caused by traffic. Studies showed positive results in this area, as Laib F. and others represented in [18]. At a constant speed of 20 km/h, around 25 electric buses were shown to emit noise at the level of one conventionally driven bus. It was also noted that the introduction of electric buses in cities with high traffic has a limited impact on noise reduction due to their low percentage share of public transport. An analysis of the results of the studies found that reducing noise emissions by using electric buses in the city had the greatest benefit when there was a large share of buses in the total traffic and when buses moved at low speeds (up to 50 km/h) in a city where there is a small percentage of additional heavy traffic. Another positive aspect of the introduction of electric buses is energy saving. Nurhadi L. and others [19] pointed out that the electric propulsion system was the most energy-efficient and sustainable choice. The authors also pointed out that existing diesel buses were one of the worst solutions in terms of both energy efficiency and cost, and by far the worst choice in terms of emissions throughout the life cycle. The authors also predict that, compared to a diesel bus, the total cost of ownership over an 8-year period for an electric bus, would be 20% lower.

### *1.2. The Sustainable Development of the City*

A city's increasing environmental pollution requires the implementation of innovative solutions for the sustainable development of the city, especially in the transport sector. The construction of ring roads, the development of environmentally friendly, public transport for passengers as well as the use of alternative energy sources in the supply of vehicles are solutions that will significantly achieve the objectives aimed at sustainable urban development.

In the context of sustainable urban development, Sustainable Development Goals (SDGs) are designed to educate the public, mobilize, and empower all city actors to deal with practical problem-solving and address the specific challenges of poverty and access to infrastructure [20]. It is also important to promote integrated and innovative urban infrastructure design and to promote effective spatial planning. Spatial planning has a significant impact on reducing the negative impact of the use of urban areas. In 2007, New York introduced the concept of #OneNYC for a strong and just city, which is based on the protection of the economy and climate and includes eradicating the poverty of 800,000 people by 2025 and eliminating waste (i.e., the zero waste concept) [21]. Undoubtedly, cities are responsible for the majority of global greenhouse gas emissions, hence the need to develop a strategy for action against climate change and disasters [22]. San Francisco, Copenhagen, Sydney, and Vancouver have set a target of obtaining 100% of the city's energy from renewable sources [21].

However, attention should also be paid to the growing criticism of economic growth and the preservation of the Green Deal [23], especially in the richest countries "one in which economic growth should lose its privileged position as the touchstone of policy and institutional success" [24,25]. It is important to stress that "Although many economies around the world are indeed getting better at producing commodities more cleanly and efficiently, a process known as 'relative decoupling', the overall ecological impact is nevertheless still increasing, because every year, increasing numbers of commodities are being produced, exchanged, and consumed as a result of growing economies" [26]. When introducing goods onto the market that are "safer for the environment", we do not always act in its favor. One of the paradoxes depicted in the literature of the subject [23], is an example of more fuel-efficient vehicles, where the effect of the rebound is an increase in the number of drivers and a higher number of vehicles purchased than is the case for "less eco" vehicles. In conclusion, it is necessary to raise awareness and actions leading to sustainable development, but the effects of the so-called 'rebound', associated with the programs, should not be underestimated.

In addition, the idea of a Smart City has been met with an increasing number of skeptics of this type of solution, raising more and more questions and controversies due to the collection of data and their unrestricted access thereto, thus submitting to the techno-political ordering of society [27–30]. It is indicated that an inability to implement a Smart City solution and the misuse of data can affect public safety and violate the citizens' right to privacy, which easily does more harm than good [31]. "One important and constant characteristic of these different visions, however, is that they aim to evoke positive change and innovation—At least as its proponents see it—Via digital ICT; essentially, building an IoT—of city-like proportions—by installing networked objects throughout the urban environment (and even in human bodies) for a wide range of different purposes [29]".

This paper proposes and explores a sustainable urban development card for the Polish city of Zielona Góra, the use of which will ensure sustainable public transport. In 2018, Poland introduced a law on electro-mobility, which imposes an obligation to gradually replace bus rolling stock with zero emission buses by 2028, by a minimum of 30% for cities with over 50,000 residents. The implementation of this solution is aimed at improving the quality of life of residents and reducing greenhouse gas emissions. In Poland, the response to the adopted UN Sustainable Development Goals, which have become a new development map for the world, is the "Sustainable Cities" Program. It is assumed that activities in the field of sustainable urban development, economic growth, and environmental protection are to be undertaken to achieve SDGs by 2030 [22]. Deep-seated changes in Polish cities will necessitate the protection of the environment with simultaneous economic growth. Such assumptions require the implementation of appropriate strategies in many areas of the city's functioning. One of the cities implementing the "Sustainable Cities" program is Kielce. The flagship idea is the Smart City Platform, which is to provide access to data, its processing, analysis, visualization and sharing, in order to provide public e-services [32]. The City of Kielce also implements a low-emissions economy plan and conducts integrated management of municipal property.

Activities for a low-emissions economy constitute a common challenge for Polish cities because "in the centers of large cities, some 60% of air pollution comes from transport, according to estimates, while in Warsaw, it is as much as 63%; furthermore, six Polish cities are among the most polluted in Europe" [33]. In Poland, 198 electric buses were registered in 2019 as part of a low-emission transport strategy [34]. In July 2019, Zielona Góra had the largest fleet of electric buses in Poland. Therefore, further analysis was based on the case study of the city of Zielona Góra, in order to show a good example of the impact on sustainable urban development with the implementation of electric buses in Polish cities with over 140,000 inhabitants.

The authors have focused their research on analyzing the effectiveness of implementing electric buses in the city of Zielona Góra and on conducting a research survey among passengers with regard to satisfaction with the use of these buses. In order to solve this, we developed sustainable indicators (SD) directly affecting the three main aspects of SD for public transport, *viz.*, environmental, social, and economic dimensions. Therefore, in this work, the two main areas of an analysis of the parameters of a system of low carbon energy public transport were formulated: (1) Sustainable Development Goals (SDGs) and (2) indicators of the Satisfaction Rate of Public Transport Passengers (SPTP). This last proposed solution is an analysis of electric bus parameters and passenger satisfaction in terms of sustainable urban development. Moreover, no research analysis has been carried out on sustainable urban development the impact of the implementation of electric buses in Polish cities with over 140,000 inhabitants.

Based on the analysis carried out, the need was pointed out to create a new approach to analyzing the system of low carbon energy public transport, combining SDGs and SPTP, the so called sustainable urban development card. Section 2 describes the research materials and methods while Section 3 gives details of the results of the research. Finally, a discussion is provided and the direction of further work is presented.

## 2. Materials and Methods

### *2.1. Data Collection*

The results of the analysis of the level of sustainable urban development in Poland that was published by Arcadis [35] showed that Poland is still a country with very low sustainable urban development. In order to build the sustainable urban development card, the parameters for the analysis of a system of low carbon energy public transport were formulated, based on the opinions of experts involved in the implementation of the "Sustainable Cities" Program in Poland in two areas: (1) Sustainable Development Goals (SDGs) and (2) the indicators of the Satisfaction Rate of Public Transport Passengers (SPTP). Therefore the studies were carried out in two stages: (1) SDG measurement and (2) main survey studies.

#### 2.1.1. Sustainable Development Goal (SDG) Measurement

The following Sustainable Development Goals (SDGs) were defined:

- $SDG_e$—Environmental: $SDG_{e1}$—reduced energy consumption; $SDG_{e2}$—reduced environmental pollution; $SDG_{e3}$—increase in benefits to society without compromising future generations.
- $SDG_s$—Social: $SDG_{s1}$—customer satisfaction; $SDG_{s2}$—increasing innovation; $SDG_{s3}$—comfort of residents
- $SDG_c$—Economic: $SDG_{c1}$—price.

The data for calculating the indicated SDG were obtained from the Municipal Department of Transport in the city of Zielona Góra. The energy consumption index was the daily average of kwH/km consumption, which was adopted on the basis of three periods: the last quarters of 2018, 2019, 2020, and August. The remaining SDG values were adopted according to data for August 2020.

#### 2.1.2. Main Survey Studies

The main survey research was carried out at stops and in public transport vehicles on a representative sample of passengers aged 16–75 [$n$ = 1022], with an estimated gender and age structure, corresponding to the entire population of the inhabitants of Zielona Góra; the research was conducted in 2019. At least 700 interviews were conducted on electric vehicles and at least 300 were conducted on vehicles with internal combustion engines. The data were collected using a direct questionnaire interview (PAPI). The survey questionnaire (Appendix A) had ten questions: nine closed and one open, in which it was possible to provide interviewees with possible comments of their own, regarding the Green Mountain and the public transport of electric buses, newly put into operation. Information on the gender structure of the respondents is presented in Figure 1, while the age of the respondents is shown in Figure 2.

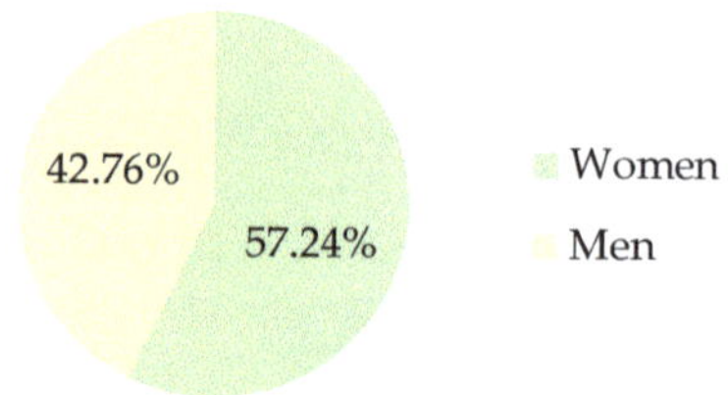


**Figure 1.** The gender structure of the respondents.

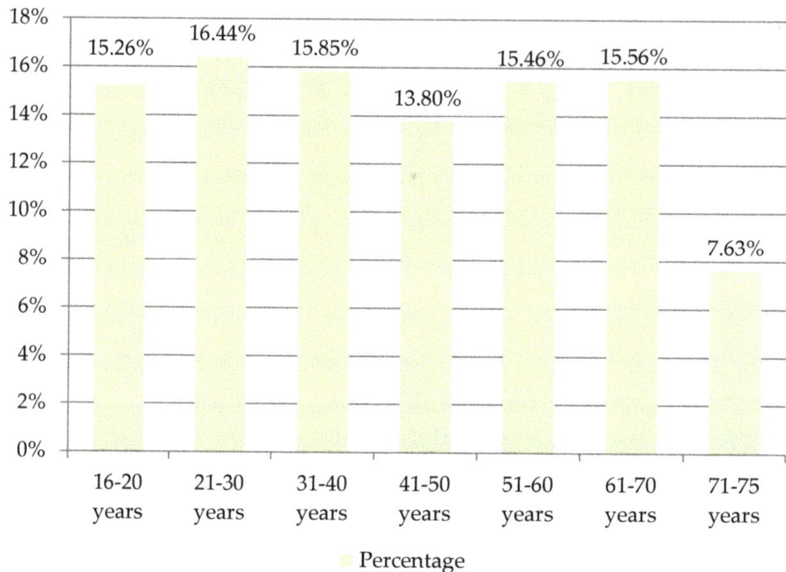


**Figure 2.** The age of the respondents in %.

The survey asked respondents to identify their age group (aged 16–20, 21–30, 31–40, 41–50, 51–60, 61–70, and 71–75). The survey involved 585 women, representing 57.24% of the total number of respondents, of which the percentage was women: 54.49% aged 16–20, 49.4% aged 21–30, 54.94% aged 31–40 years, 62.41% aged 41–50 years, 59.49% aged 51–60 years, 62.26% aged 61–70 years, and 60.26% aged 71–75. For the opposite sex, 437 men participated in the survey, representing 42.76% of all respondents, of which 45.51% were men aged 16–20, 50.6% aged 21–30 years, 45.06% 31–30 years, 37.59% aged 41–50 years, 40.51% aged 51–60 years, 37.74% aged 61–70%, and 39.74% aged 71–75 years.

Indicators of the satisfaction rate of Public Transport Passengers (SPTP) were defined in the following way:

Factors of SPTP ($SPTP_1$–$SPTP_4$) were based on feedback surveys and their sources are listed here: SPTP: *"I know the advantages of electric buses in my city"*: $STPT_1$—"Keep Calm" Electric Vehicle, $STPT_2$—smooth ride, $STPT_3$—environmentally friendly: ecology, $STPT_4$—environmentally friendly: air quality is: factor0: not very important/factor1: very important.

Factors for the evaluation of electric buses by passengers of public transport vehicles were based on feedback surveys and their sources are listed here: Evaluation of electric buses: the degree to which a city creates innovative solutions for sustainable urban development, using the scale: factor1: not very important; factor2: not important; factor3: quite important; factor4: important; factor5: very important.

### 2.2. *Case Study: The City of Zielona Góra, Poland*

As above-mentioned, the city of Zielona Góra was chosen for further analysis since it has the largest fleet of electric buses in Poland. The city of Zielona Góra, together with the Municipal Department of Transport (MZK) has completed a project entitled "The integrated low-emission public transport system in Zielona Góra". The city of Zielona Góra is the largest city in Lubuskie Province with about 140,000 inhabitants. The project was implemented from the operational program Infrastructure and Environment 2014–2020 action 6.1. The development of public transport in cities was co-financed by the European Union budget and its value was PLN 284,825,840.94. The project proposed the following components [36]:

- purchase of new, low-floor electric buses (12 m): 43 units;
- purchase of new low-floor diesel buses (18 m): 17 units;
- battery charging infrastructure;
- modernization of the Municipal Department of Transport bus depot;
- construction of a passenger transport interchange;
- canopy at the railway station;
- modernization of the viaduct under the railway tracks, on ul. Batory;
- construction and reconstruction of loops and stops with accompanying infrastructure; and
- modernization of the real-time passenger information system and the MZK management system.

The city of Zielona Góra has been the only city in Poland to replace weekend transport lines with 100% electric buses. In addition, mechanics and drivers from MZK indicate a low failure rate of new vehicles when compared with earlier, complex diesel engines. In total, 11 fast charging stations with a total number of 32 charging stations and one charging station at the depot were established in Zielona Góra. In addition to the fast charging station, there are 25, two-station, free chargers in the city. The fast charging time of one bus is 20 min. A full charge is enough to travel 50 km.

### 2.3. *Research Model*

The research model (Figure 3) posits, from the preceding argument, that the implementation of electric buses increases the level of sustainable urban development. The main objective at the local level, in terms of transport, is to increase the safety of traffic users and the quality of life of residents and to reduce environmental damage. Our study focused on the actions taken by the city of Zielona Góra, in order to ensure sustainable public transport.

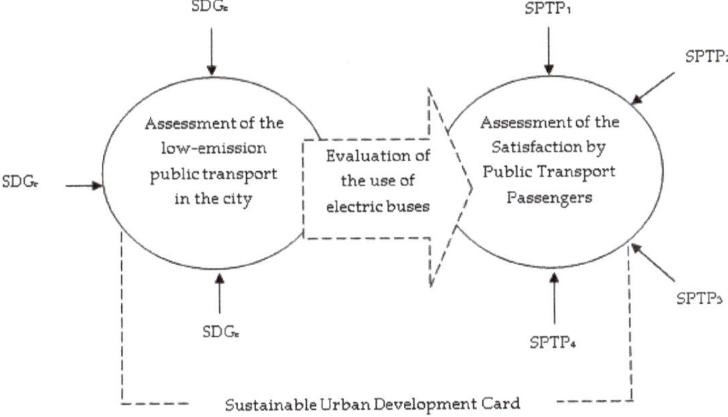


**Figure 3.** Research model.

### 2.4. Analytic Method

The research model (Figure 3) was analyzed using the correlation and further regression approach in order to build the sustainable urban development card. A moderated correlation approach, using Statistica version 13.3 (StatSoft Polska Sp. z o.o., Kraków, Poland), was used.

Finally, we proposed the sustainable urban development card (Table 1). The level of sustainable urban development is defined as: 1: unacceptable; 2: low; 3: good; 4: very good; and 5: excellent, based on the values of each of the SDGs obtained. The level of relationships between the Satisfaction Rate of Public Transport Passengers and the advantage of electric buses is also defined as 1: unacceptable; 2: low; 3: good; 4: very good; and 5 as excellent, based on the obtained results of the correlation analysis.

**Table 1.** Sustainable urban development card.

| Sustainable Urban Development | Level | Results |
|---|---|---|
| $SDG_e$ | $\frac{\text{level(SDGe1)+level(SDGe2)+level (SDGe3)}}{3}$ | 1 or 2 or 3 or 4 or 5 |
| $SDG_s$ | $\frac{\text{level(SDGs1)+level(SDGs2)+level (SDGs3)}}{3}$ | 1 or 2 or 3 or 4 or 5 |
| $SDG_c$ | level(SDGc1) | 1 or 2 or 3 or 4 or 5 |
| Evaluation of the use of electric buses (E) | $\frac{\text{sum of passenger response levels (scale: 1 to 5)}}{\text{number of responses}}$ | 1 or 2 or 3 or 4 or 5 |
| $STPT_1$ | Value of correlation analysis | 1 or 2 or 3 or 4 or 5 |
| $STPT_2$ | Value of correlation analysis | 1 or 2 or 3 or 4 or 5 |
| $STPT_3$ | Value of correlation analysis | 1 or 2 or 3 or 4 or 5 |
| $STPT_4$ | Value of correlation analysis | 1 or 2 or 3 or 4 or 5 |

This article offers an assessment of low-emission public transport and, in addition, an explanation of the influence of the implementation of electric buses on the increased SPTP. On the basis of the current analysis of the literature on the subject and of SDGs defined, in the context of sustainable urban development and the results of the research survey, the approach for the analysis of a system of low carbon energy public transport combining SDGs and SPTP, was defined (Figure 1). A questionnaire was developed, based on which, SPTP studies on a representative sample of passengers of public transport vehicles in Zielona Góra, were conducted. The surveys applied tested the research model (Figure 1) and were developed by defining scales to fit the meaning of the codification of the Satisfaction Rate of Public Transport Passengers.

## 3. Research Results

Based on the data received from the Municipal Department of Transport, Zielona Góra—and in order to build the Sustainable Urban Development Card for that city—the values of SDGs were obtained (Table 2). The energy consumption index is the daily average of kwH/km consumption, which was adopted on the basis of three periods: the last quarter of 2018, 2019, 2020, and August. The remaining indicator values were adopted according to data for August 2020.

**Table 2.** Values of Sustainable Development Goals (SDGs) for the city of Zielona Góra.

| SDGs | Value | Level |
|---|---|---|
| $SDG_{e1}$ | 1.23 kwH/km (average daily consumption kwH/km) | 4 |
| $SDG_{e2}$ | 0 GHG | 5 |
| $SDG_{e3}$ | Reducing risk management related to the climate | 4 |
| $SDGs_1$ | 89% | 4 |
| $SDGs_2$ | Lithium batteries, capacity 175 kwH | 4 |
| $SDGs_3$ | Noise [dB] = 63 | 4 |
| $SDGc_1$ | One-off ticket price [Polish]:<br>Paper ticket 1.50–3.00<br>Electronic wallet 1.00–2.60 | 5 |

The level of sustainable urban development for each SDG was defined, based on a comparison with conventional diesel buses (Table 3).

**Table 3.** Comparison of the value of SDGs in the city of Zielona Góra between electric buses and diesel buses.

| SDGs | Value: Electric Buses | Value: Buses with a Diesel Engine |
|---|---|---|
| $SDG_{e1}$ | 1.23 kwH/km (average daily consumption kwH/km) | 45l/100 km for 12 m buses, 50l/100 km for articulated buses |
| $SDG_{e2}$ | 0 GHG | 924 g $CO_2$ eq/poj-km [31] |
| $SDG_{e3}$ | Reducing risk management related to the climate | - |
| $SDGs_1$ | 89% | - |
| $SDGs_2$ | Lithium batteries, capacity 175 kwH | E-card |
| $SDGs_3$ | Noise [dB] = 63 | Noise [dB] = 80 |
| $SDGc_1$ | One-off ticket price [Polish]:<br>Paper ticket 1.50–3.00<br>Electronic wallet 1.00–2.60 | One-off ticket price [Polish]:<br>Paper ticket 1.50–3.00<br>Electronic wallet 1.00–2.60 |

The electricity used in the city for charging electric buses quickly is taken from Enea, while the electricity for charging at the depot is taken from the Combined Heat and Power Plant (gas). Compared to diesel prices, the electricity needed to drive 100 km on an eco-friendly bus is now half the price [37].

The next part of our sustainable urban development card for Zielona Góra is an evaluation of electric buses by passengers of public transport vehicles. The value was based on feedback surveys from a representative sample of passengers aged 16–75 [$n = 1022$] and was conducted in 2019. The level of the evaluation of the use of electric buses is at 4 (Equation (1)).

$$E = \frac{4492}{1022} = 4.39 \tag{1}$$

Table 4 presents descriptive correlations for the main variables included in the research model (Figure 3) using Statistica ver.13.3. The data were carefully examined with respect to linearity, equality of variance, and normality. No significant deviations were detected.

**Table 4.** Correlation analysis.

| Relations | Correlation | r2 | t | p |
|---|---|---|---|---|
| $STPT_1$—"Keep Calm" Electric Vehicle/Satisfaction Public Transport Passengers | 0.4239 | 0.1797 | 14.9486 | 0.0000 |
| $STPT_2$—smooth ride/Satisfaction Rate of Public Transport Passengers | 0.0318 | 0.0010 | 1.0172 | 0.3093 |
| $STPT_3$—environmentally friendly: ecology/Satisfaction Public Transport Passengers | 0.0027 | 0.0000 | 0.0886 | 0.9294 |
| $STPT_4$—environmentally friendly: air quality/Satisfaction Rate of Public Transport Passengers | −0.0185 | 0.0003 | −0.5908 | 0.5547 |

where r2 is the coefficient of determination; T is the value of t statistics examining the significance of the correlation coefficient; and $p$ is the probability value.

The analysis showed one significant relationship between the advantages of using electric buses in the city: "Keep Calm" and the satisfaction of the passengers using public transport. Unfortunately, no significant relationship was found between $STPT_2$, $STPT_3$, $STPT_4$ and the increased satisfaction of public transport passengers. Therefore, to define the nature of significant interactions of the influence of the advantages of using electric buses on the increase in SPTP—in the context of a Polish city—the study tested the hypotheses using regression analyses that estimate this impact (Equation (2)).

$$SPTP = 0.6736 + 0.2978x\text{STPT1} \tag{2}$$

Based on the analysis, it was noticed that the advantages of using electric buses "Keep Calm" increased the satisfaction of public transport passengers (2, Figure 4).

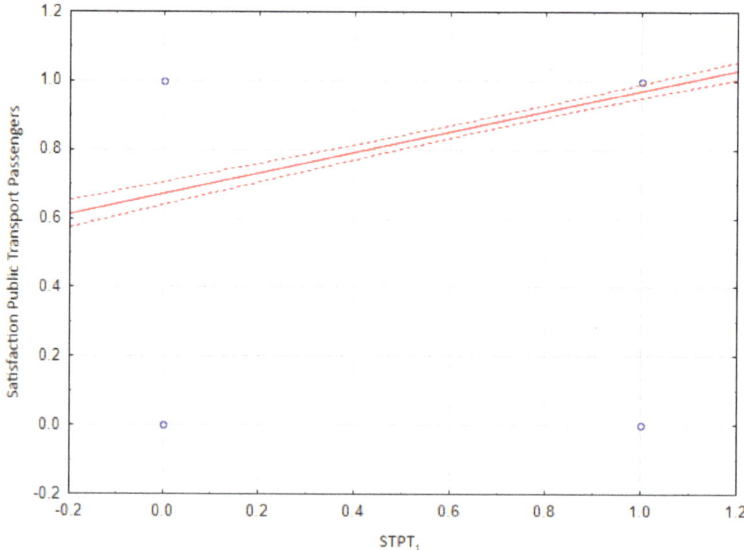


**Figure 4.** Sustainable urban development card.

The statistical results showed that the one advantage examined in using electric buses was the satisfaction of public transport passengers. Finally, Table 5 presents the sustainable urban development card for Zielona Góra, based on the research results in Tables 2 and 4.

**Table 5.** Sustainable urban development card for a Polish city.

| Sustainable Urban Development | Level | Results |
|---|---|---|
| $SDG_e$ | 4.33 | 4 |
| $SDG_s$ | 4 | 4 |
| $SDG_c$ | 5 | 5 |
| E | 4.39 | 4 |
| $STPT_1$ | 0.4239 | 4 |
| $STPT_2$ | 0.0318 | 1 |
| $STPT_3$ | 0.0027 | 1 |
| $STPT_4$ | −0.0185 | 1 |

It is clearly obvious that further actions should be applied in increasing the satisfaction of public transport passengers and increasing awareness of them in sustainable urban development. It seems that for the passengers of Zielona Góra, the following factors, namely, a smooth and environmentally friendly ride as opposed to ecology and air quality, are not important in the context of increased comfort.

The proposed sustainable urban development card can be applied to all cities implementing a revision of public transport rolling stock as part of sustainable development. SDGs and STPTs can be expanded or modified in terms of the specific needs of the cities concerned. The proposed solution enables the transport authorities in the city to control undertakings in real time, reduce costs, save resources and energy, implement an environmental policy, and constitutes a database for further actions.

## 4. Discussion

This work compared the adopted SDGs for electric buses with transit buses. Electric buses have allowed petroleum-based measures for electricity to be completely eliminated. It is important to point out that electricity comes from both the Enea energy distributor and natural gas, which is an alternative source of energy. Another parameter included in the analysis was greenhouse gas emissions, which for transit buses is 924 g $CO_2$ eq/poj-km, while for electric buses, it is 0 g $CO_2$ eq/poj-km. It should be stressed, however, that GHG emissions only concern vehicle operation with the possible emissions from additional equipment, machinery, or passengers not being taken into account. The indicator studied, which in the literature of the subject arouses much controversy, was noise. According to the analysis, the noise produced by the new low-emission bus fleet decreased by 17 dB, which positively affects the comfort of life of residents, especially at night, when traffic is very limited. As indicated in the analysis of the literature, the reduction of noise emissions by public transport, during peak hours in medium and large cities, has little bearing on the health and comfort of the population due to the low percentage of traffic. Another aspect related to noise is safety. This parameter was not included in the survey, prompting the authors to examine this dependency in subsequent surveys. An important summary of the results is the overall assessment of the new low-emission public transport fleet by passengers of the MZK in Zielona Góra. MZK received a rating of 4.39, on a scale of 1 to 5, which is a very good result and bodes well for the further achievement of SDGs in the city.

In our example, we created a sustainable urban development card in order to monitor improvements in the state of the atmosphere through the sustainable and efficient use of energy carriers [37]. Due to the changes introduced in 2019, the Municipal Transport Department in Zielona Góra commissioned a study on the quality of transport services concerning the evaluation of new rolling stock by passengers. The Municipal Transport Department in Zielona Góra has also introduced changes to the availability and price of tickets that are beneficial for passengers. Using an electronic wallet allows for cheaper travel than a paper ticket. In order to facilitate the purchase of a travel ticket, the passenger can use several forms: e-card, electronic wallet, paper wallet, application (app).

Bus tickets have also been used in card readers, which allows you to pay for the ticket with a bank card or smartphone. Today, residents of Zielona Góra can benefit from the facilities introduced including the transfer center at the railway station, low-emission and silent electric buses, and an extensive, passenger information system. The fleet of buses implemented is passenger-friendly and, importantly, four times quieter than diesel buses as well as cheaper to maintain and are much greener.

In Zielona Góra, the low-emission public transport project will continue to be developed. A contract has been signed for another 12 electric buses; ultimately the city wants to switch to 100% electric buses and give up conventional buses altogether. Bus manufacturer URSUS is also pledging changes to further reduce its environmental impact by increasing battery capacity from 175 kWh to 250 kWh [38]. The solutions introduced are part of the sustainable development strategy of Zielona Góra. In addition to reducing environmental pollutants, the introduction of electric buses has reduced noise emissions. The average noise emission of an electric bus during a stop-over is about 64 dB, while for the diesel-powered bus, it is 80 dB. Another aspect worth paying attention to is the use of gas as an energy source for charging electric buses. In terms of the work of electric buses in Zielona Góra, it should be emphasized that [39]:

- buses connect 430 times a day in order to charge up;
- on average, a bus charges up to 23 kwH and a maximum of 68 kwH of energy on a single charge;
- the longest route between charging is 43 km and charging time, after this mileage, is 17 min as a maximum; and
- the annual electricity consumption of MZK buses in 2019, amounted to more than 2400 mwH.

## 5. Conclusions

The sustainable urban development card presented can be a good benchmark as to how cities can improve their level of sustainability. The idea, as presented, is an innovative approach to monitoring and analyzing low carbon energy public transport. The issue of SDGs in countries and cities is of key importance on a global scale. The new approach combines the SDGs, STPTs, and E as well as the research methods of survey and correlation analysis.

Thanks to the implementation of our card, it is possible:

- to present a level of sustainable urban development;
- to evaluate the adopted SDGs, STPTs, and E;
- to obtain a table defining the level of the Sustainable Urban Development, enabling the recommended corrective actions to be determined; and
- to constantly monitor the corrective actions implemented in the city.

As with all studies, this study had certain limitations, which further research should be able to overcome. First, the application of a model was shown in the example of the Polish city investigated and all the indicators were measured at the same time; it would, therefore be useful to provide such research over a longer period of time. Second, it should be borne in mind that an increase in energy efficiency may lead to fewer energy savings than would be expected by simply multiplying the change in energy efficiency by the energy use *prior to the change* [40]. These conclusions and limitations suggest proposals for the direction of future research, especially in the context of extending the proposed approach to include the aspect of measuring the so-called "rebound effect".

**Author Contributions:** Conceptualization, J.P.-M.; Methodology, J.P.-M. and H.Ł.; Software, J.P.-M. and H.Ł.; Validation, J.P.-M.; Formal analysis, J.P.-M. and H.Ł.; Investigation, J.P.-M.; resources, J.P.-M. and H.Ł.; data curation, J.P.-M. and H.Ł. writing—original draft preparation, J.P.-M. and H.Ł.; Writing—review and editing, J.P.-M. and H.Ł.; Visualization, J.P.-M. and H.Ł.; Supervision, J.P.-M.; Project administration, J.P.-M.; Funding acquisition: J.P.-M. and H.Ł. All authors have read and agreed to the published version of the manuscript.

**Funding:** This research received no external funding.

**Conflicts of Interest:** The authors declare that there are no conflict of interest.

## Appendix A

**Table A1.** The Survey Questionnaire.

| No | Questions | Answer |
|---|---|---|
| 1. | How do you evaluate the public transport in Zielona Góra in general, with particular reference to the services of MZK Zielona Góra)? | (a) Very good<br>(b) Good<br>(c) Adequate<br>(d) Inadequate, because . . . .<br>(e) I have no opinion |
| 2. | Are you satisfied with the introduction of electric buses in Zielona Góra? | (a) Very good<br>(b) Good<br>(c) Adequate<br>(d) Inadequate, because . . . .<br>(e) I have no opinion |
| 3. | Have you already travelled on the new electric buses? | (a) Yes<br>(b) No<br>(c) I have no opinion |
| 4. | What do you think of these buses? | (a) Very good<br>(b) Good<br>(c) Adequate<br>(d) Inadequate, because . . . .<br>(e) I have no opinion |
| 5. | What do you think is the greatest advantage of electric buses in my city; please choose one advantage | (a) "Keep Calm" Electric Vehicle,<br>(b) Smooth Ride,<br>(c) Environmentally Friendly: ecology,<br>(d) Environmentally friendly: air quality<br>(e) I have no opinion |
| 6. | Do you think the introduction of electric buses is important for creating innovative solutions for sustainable urban development? | (a) Very important<br>(b) Important<br>(c) Quite important<br>(d) Not important<br>(e) Inadequate<br>(f) I have no opinion |
| 7. | Do you think the introduction of electric buses has had a significant impact on air quality in the city? | (a) Yes<br>(b) No, because . . . .<br>(c) I don't know |
| 8. | Should the city of Zielona Góra continue to invest in electric buses? | (a) Yes<br>(b) No, because . . .<br>(c) I don't know |

## References


1. International Energy Agency. Available online: https://www.iea.org/ (accessed on 18 October 2020).
2. Hashmi, M.; Hanninen, S.; Maki, L. Survey of smart grid concepts, architectures, and technological demonstrations worldwide. In Proceedings of the IEEE PES Innovative Smart Grid Technology Latin America (ISGT LA), Medellin, Colombia, 19–21 October 2011; IEEE: Piscataway, NJ, USA, 2011; pp. 1–7.
3. Saboori, H.; Mohammadi, M.; Taghe, R. Virtual Power Plant (VPP), definition, concept, components and types. In Proceedings of the Asia-Pacific Power Energy Engineering Conference, Wuhan, China, 25–28 March 2011; IEEE: Piscataway, NJ, USA, 2011.
4. Tahmasebi, M.; Pasupuleti, J. Self-scheduling of wind power generation with direct load control demand response as a virtual. *Indian J. Sci. Technol.* **2013**, *6*, 5443–5449. [CrossRef]
5. Seyfang, G.; Park, J.J.; Smith, A. A thousand flowers blooming? An examination of community energy in the UK. *Energy Policy* **2013**, *61*, 977–989. [CrossRef]

6. Klein, S.J.W.; Coffey, S. Building a sustainable energy future, one community at a time. *Renew. Sustain. Energy Rev.* **2016**, *60*, 867–880. [CrossRef]
7. Summeren, L.; Wieczorek, A.; Bombaerts, G.J.; Verbong, G. Community energy meets smart grids: Reviewing goals, structure, and roles in Virtual Power Plants in Ireland, Belgium and the Netherlands. *Energy Res. Soc. Sci.* **2020**, *63*, 101415. [CrossRef]
8. Pudjianto, D.; Ramsay, C.; Strbac, G. Virtual power plant and system integration of distributed energy resources. *Renew. Power Gener. IET* **2007**, *1*, 10–16. [CrossRef]
9. Pihlatie, M.; Kukkonen, S.; Halmeaho, T.; Karvonen, V.; Nylund, N.O. Fully electric city buses-the viable option. In Proceedings of the IEEE International Electric Vehicle Conference (IEVC), Florence, Italy, 17–19 December 2014; IEEE: Piscataway, NJ, USA, 2014.
10. Asamer, J.; Graser, A.; Heilmann, B.; Ruthmair, M. Sensitivity analysis for energy demand estimation of electric vehicles. *Transp. Res. Part D Transp. Environ.* **2016**, *46*, 182–199. [CrossRef]
11. Wang, J.; Besselink, I.J.M.; Nijmeijer, H. Battery electric vehicle energy consumption modelling for range estimation. *Int. J. Electr. Hybrid Veh.* **2017**, *9*, 79. [CrossRef]
12. Vepsäläinen, J.; Kivekäs, K.; Otto, K.; Lajunen, A.; Tammi, K. Development and validation of energy demand uncertainty model for electric city buses. *Transp. Res. Part D Transp. Environ.* **2018**, *63*, 347–361. [CrossRef]
13. Shankar, R.; Marco, J. Method for estimating the energy consumption of electric vehicles and plug-in hybrid electric vehicles under real-world driving conditions. *IET Intell. Transp. Syst.* **2013**, *7*, 138–150. [CrossRef]
14. Beckers, C.; Besselink, I.; Nijmeijer, H. Assessing the impact of cornering losses on the energy consumption of electric city buses. *Transp. Res. Part D* **2020**, *86*, 102360. [CrossRef]
15. Woodcock, A.; Topolavic, S.; Osmond, J. The multi-faceted public transport problems revealed by a small scale transport study. In *Advances in Human Aspects of Road and Rail Transportation*; Stanton, N.A., Ed.; CRC Press: San Francisco, CA, USA, 2013; pp. 321–331.
16. Jwa, K.; Lim, O. Comparative life cycle assessment of lithium-ion battery electric bus and Diesel bus from well to Wheel. *Energy Procedia* **2018**, *145*, 223–227. [CrossRef]
17. Liu, Q.; Hallquist, A.M.; Fallgren, H.; Jerksjo, M.; Jutterström, S.; Salberg, H.; Hallquist, M.; Le Breton, M.; Pei, X.; Pathak, R.K.; et al. Roadside assessment of a modern city bus fleet: Gaseous and particle emissions. *Atmos. Environ. X* **2019**, *3*, 10044. [CrossRef]
18. Laib, F. Modelling noise reductions using electric buses in urban traffic. A case study from Stuttgart, Germany. *Transp. Res. Procedia* **2019**, *37*, 377–384. [CrossRef]
19. Nurhadi, L.; Boren, S.; Ny, H. A sensitivity analysis of total cost of ownership for electric public bus transport systems in Swedish medium sized cities. *Transp. Res. Procedia* **2014**, *3*, 818–827. [CrossRef]
20. Urban SDG. Available online: http://urbansdg.org/about/ (accessed on 22 September 2020).
21. Local Governments for Sustainability. Available online: https://www.local2030.org/library/234/ICLEI-SDGs-Briefing-Sheets-04-The-importance-of-all-Sustainable-Development-Goals-SDGs-for-cities-and-communities.pdf (accessed on 19 September 2020).
22. UNGC. Available online: https://ungc.org.pl/programy/zrownowazone-miasta/ (accessed on 21 September 2020).
23. Alexander, S. Planned economic contraction: The emerging case for degrowth. *Environ. Politics* **2012**, *21*, 349–368. [CrossRef]
24. Stiglitz, J.; Sen, A.; Fitoussi, J.P. *Mis-Measuring Our Lives: Why GDP Doesn't Add Up*; The New Press: New York, NY, USA, 2010.
25. Property beyond Growth: Toward a Politics of Voluntary Simplicity. Available online: www.simplicityinstitute.org/publications (accessed on 19 October 2020).
26. Jackson, T. *Prosperity without Growth: Economics for A Finite Planet*; Earthscan: London, UK, 2009.
27. Angelidou, M. Smart cities: A conjuncture of four forces. *Cities* **2015**, *47*, 95–106. [CrossRef]
28. Kitchin, R. Making sense of smart cities: Addressing present shortcomings. *Camb. J. Reg. Econ. Soc.* **2015**, *8*, 131–136. [CrossRef]
29. Sadowski, J.; Pasquale, F. The Spectrum of Control: A Social Theory of the Smart City. First Monday 2015, 20. Available online: https://firstmonday.org/ojs/index.php/fm/article/view/5903/4660 (accessed on 22 September 2020).
30. Zuboff, S. *The Age of Surveillance Capitalism: The Fight for a Human Future at the New Frontier of Power*; Profile Books: London, UK, 2019.
31. Citron, D.K.; Pasquale, F. Network accountability for the domestic intelligence apparatus. *Hastings Law J.* **2011**, *62*, 1441–1494.

32. Almine. Available online: https://almine.pl/smart_city_przyklady_polska/ (accessed on 22 September 2020).
33. UN Global Compact. Zrównoważone Miasta. Życie w Zdrowej Atmosferze. 2016. Available online: https://ungc.org.pl/wp-content/uploads/2016/10/Raport-Zr%C3%B3wnowa%C5%BCone-miasta.pdf (accessed on 20 September 2020).
34. InfoBus. Available online: http://infobus.pl/polski-rynek-autobusow-elektrycznych-lipiec-2019_more_117202.html (accessed on 23 September 2020).
35. Ranking of Sustainable Polish Cities, Arcadis, Warsaw, Poland. 2018. Available online: https://www.arcadis.com/media/6/0/D/%7B60DA8546-A735-430A-BF9A-6114B0362FD7%7DRanking%20Polskich%20Miast%20Zrownowazonych%20Arcadis%20FINAL.pdf (accessed on 19 September 2020).
36. Langner, B.; Newelski, J. Integreted System of a Public Transport System which Uses Low-Carbon Energy: (in Polish). In Proceedings of the MZK Conference, Zielona Góra, Poland, 30 September 2020.
37. The Methodology for Assessing Greenhouse Gas Emissions, Warsaw 2015. Available online: https://bip.zielonagora.pl/system/obj/65592_RAPORT_cz.1__-_programy_strategie.pdf (accessed on 22 September 2020).
38. Subdisy Map. Available online: https://mapadotacji.gov.pl/projekty/746900/ (accessed on 1 September 2020).
39. Gorzow. Available online: https://gorzow.tvp.pl/48962949/2-lata-elektrycznych-autobusow-w-zielonej-gorze-jak-sie-sprawdzaja (accessed on 2 September 2020).
40. Gillingham, K.; Rapson, D.; Wagner, G. The Rebound Effect and Energy Efficiency Policy. *Rev. Environ. Econ. Policy* **2015**, *10*, 68–88. [CrossRef]

*Article*

# Smart Energy in a Smart City: Utopia or Reality? Evidence from Poland

**Aleksandra Lewandowska *, Justyna Chodkowska-Miszczuk, Krzysztof Rogatka and Tomasz Starczewski**

Department of Urban and Regional Development Studies, Faculty of Earth Sciences and Spatial Management, Nicolaus Copernicus University, 87-100 Toruń, Poland; jchodkow@umk.pl (J.C.-M.); krogatka@umk.pl (K.R.); tomaszstarczewski1@wp.pl (T.S.)

* Correspondence: amal@umk.pl; Tel.: +48-56-611-2633

Received: 30 September 2020; Accepted: 3 November 2020; Published: 5 November 2020

**Abstract:** The main principles of the smart city concept rely on modern, environmentally friendly technologies. One manifestation of the smart city concept is investments in renewable energy sources (RES), which are currently a popular direction in urban transformation. It makes sense, therefore, to analyse how Polish cities are coping with this challenge and whether they are including the implementation of RES facilities in their development strategies. The aim of the article is to analyze and assess the level at which renewable energy facilities are being implemented or developed in the urban space of cities in Poland as a pillar of the implementation of the smart city concept. This goal is realized on two levels: the theoretical (analysis of strategic documents) and the practical (analysis of the capacity of RES installations, questionnaire studies). The study shows that renewable energy installations are an important part of the development strategies of Polish cities, and especially of those that aspire to be termed "smart cities". Moreover, it is shown that the predominant RES facilities are those based on solar energy.

**Keywords:** smart city; renewable energy; distributed generation; urban policy; Poland

---

## 1. Introduction

Eliminating unfavourable aspects of urban management and urban living, showing a city makes energy savings and providing it with a good metabolism—these are all markers of the implementation of the smart city concept [1–6]. The smart city concept combines several elements, including: innovative use of technology, efficient transportation, sustainable energy consumption, and a clean environment [7–10]. All these components are intended to improve residents' quality of life, but also to positively affect the environment. Technological progress and socio-economic development—which are clearly visible in large urban centres in particular—increase energy demand. Modern cities are therefore currently faced with the challenge of securing energy supplies, on which are founded not only the functioning of the economy as a whole, but also residents' quality of life. The key to solving emerging problems is the concept of the "smart city", which, based on data and technology, supports economic development and improves quality of life while promoting sustainable development in cities [11].

The smart city is a complex concept with various definitions [2,9,12]. The idea features two main threads. The first is related to the use of Information and Communication Technologies (ICT) and modern urban technologies—these are mainly technical solutions [13–16]. The second is related to the role of human capital in the development of a smart city [17,18]. Furthermore, the smart city can be taken as a concept for urban development within specific fields of activity. These are usually assumed to be: the economy, residents, development management, mobility, the environment, energy, and quality of life [19]. One can analyse individual fields of activity or a specific aspect of the complex concept, but only a holistic approach creates a smart city [20].

Energy infrastructure stands out as one of the key elements of a smart city, especially because its state and structure determine whether sustainable development principles will be implemented and residents will be provided a high quality of life in a clean urban environment. Cities are increasingly important players in implementing renewable energy based on endogenous resources [21]. According to the premises of the smart city, the energy system must be wholly integrated into the local context [22]. In addition to energy demand, the potential for energy generation based on endogenous resources—locally occurring renewable energy sources (RES)—should be taken into account, with the ultimate aim of achieving independence from external fossil fuel supplies [23].

The development of RES in cities and regions is a research subject in many countries. Examples of research on RES in cities of the region are presented in Table 1.

**Table 1.** Research on renewable energy sources in cities and regions (selected).

| Region | References |
|---|---|
| Europe | Eicker (2012) [24]; Gerpott and Paukert (2013) [25]; Gabillet (2015 [26]); Kılkış (2016) [27]; Petersen (2016) [28]; Kazak, et al. (2017) [29]; Ahas, et al. (2019) [30]; Bahers, et al. (2020) [31]. |
| North America | Hammer (2008) [32]; Denis and Parker (2009) [33]; Moscovici et al. (2015) [34]; Bagheri et al. (2018) [35]; DeRolph et al. (2019) [36]; Hess and Gentry (2019) [37]; Kouhestani et al. (2019) [38]. |
| Central and South America | Ramírez et al. (2000) [39]; Huacuz (2005) [40]; De Araújo et al. (2008) [41]; Fonseca and Schlueter (2013) [42]; Cedeno et al. (2017) [43]; Pérez-Denicia et al. (2017) [44]; Lino and Ismail (2018) [45]. |
| Africa | Bugaje (2006) [46]; Cloutier and Rowley (2011) [47]; Zawilska and Brooks (2011) [48]; Gumbo (2014) [49]; Akuru et al. (2017) [50]; Bouhal el at. (2018) [51]. |
| Asia | Jebaraj and Iniyan (2006) [52]; Bilgen el al. (2008) [53]; Cheng and Hu (2010) [54]; Farooq and Kumar (2013) [55]; Schroeder and Chapman (2014) [56]; Madakam and Ramaswamy (2016) [57]; Noorollahi et al. (2017) [58]; Yuan et al. (2018) [59]; Awan (2019) [60]; Fraser (2019) [61]; Meng et al. (2019) [62]. |
| Australia | Mithraratne (2009) [63]; Martin and Rice (2012) [64]; White et al. (2013) [65]; Dowling et al. (2014) [66]; Imteaz and Ahsan (2018) [67]; Li et al. (2020) [68]. |

Source: own study.

In Poland, as in many countries, RES is being researched. Most analyses mainly involve selected case studies [69] (Table 1). However, there are no studies that, first, attempt to systematize the current state of knowledge regarding renewable energy in cities as a component of the smart city, or, second, holistically approach RES installations (small and large) to make comparisons between cities aspiring to be smart cities. This study attempts to fill precisely this gap in the current state of scientific knowledge.

Energy planning that leads to "smart" urban solutions requires that energy design be integrated with spatial and urban planning [70–73]. Planning and implementing smart city energy systems is not easy, as it involves a wide range of stakeholders—from municipal administration, through developers and energy suppliers, to current and future residents [74–77]. In addition to requiring that a synergy of trans-sectoral interests be achieved [78], the process of energy transition also appears to exhibit huge spatial inequalities [79]. In Europe, the region in need of special support for energy transformation is the central European countries (CECs), including Poland. This situation has its roots in the historical, political, and economic past. In the post-socialist countries (because it is these that we are talking about), the modernization of the energy sector is different than in the countries of western Europe. The main barriers to the transition from fossil energy to renewable sources are, on the one hand: the monoculture of conventional raw materials (i.e., of coal in Poland); the dependence on fuel imports;

and the predominantly outdated energy infrastructure; and on the other hand: the growing demand for energy conditioned by the region's socio-economic development [80].

In Polish cities, the production of electricity from RES sent to the power grid is relatively insignificant in total energy generation electricity. This situation is conditioned by three main factors. Firstly, urban areas are not inherently conducive to new large-scale energy investments, on account of their extremely limited access to free space that would class as potential locations for RES power plants. Secondly, large-scale renewable energy installations require significant financial outlays stemming from, among other things, high urban land prices and the need for complicated bureaucratic procedures. Thirdly, developing large RES installations entails significant environmental costs [81,82].

RES has the potential to be made more widespread in smart cities through the development of distributed generation (DG), which involves energy generation based on small-scale decentralized technologies that meet mainly local needs [82,83]. In accordance with the act on renewable energy sources of 20 February 2015 [84] in Poland, a small installation does not exceed 500 kW, and a micro-installation does not exceed 50 kW. Such RES installations are used both in multi-family and single-family residential buildings, as well as in autonomous power supply systems (off-grid.) Setting up micro-installations does not involve so many bureaucratic procedures, nor a license, and does not require a large space—a rare commodity in urban areas. Meanwhile, it can power various devices: street lamps, road signs, vehicles, parking meters, and such.

The aim of the article is to analyse and evaluate the level of implementation and growth of RES facilities within the urban space of cities in Poland as a pillar of the implementation of the smart city concept. The study was carried out on two levels: (1) the theoretical, involving analysis of strategic documents of Poland's largest cities, and (2) the practical/applied, involving analysis of the number and total capacity of RES installations operating in those cities, including the DG system. This study was complemented by an empirical analysis aimed at verifying the level of RES development at the local scale.

## 2. Material and Methods

Due to the multifaceted nature of the research, the authors decided to implement a multistage research procedure. To achieve the research objective, the following methods were employed:

- quantitative analyses,
- desk research on cities' strategic documents,
- a PAPI (paper and pen personal interview) survey as a case study of Bydgoszcz.

The research methods were complemented by a city walk and exploration of Bydgoszcz, with the aim of identifying renewable energy sources within the urban space and determining how they related to the concept of the "smart city" (Figure 1).

The authors began the research with a quantitative analysis of the structure of RES installations and with an analysis of the capacity of RES installations in Poland's 20 largest cities. This had a particular emphasis on micro-installations and was based on data from the Energy Regulatory Office (ERO) [85] concerning electricity producers, including small renewable energy installations. The authors decided to conduct a quantitative analysis in the 20 most populous cities in Poland, which have a high degree of socio-economic development related to the presence of renewable energy installations within the urban space (i.e., Warsaw, Kraków, Łódź, Wrocław, Poznań, Gdańsk, Szczecin, Bydgoszcz, Lublin, Białystok, Katowice, Gdynia, Częstochowa, Radom, Toruń, Sosnowiec, Rzeszów, Kielce, Gliwice, Zabrze). This stage formed the quantitative background to the research and an introduction to later considerations.

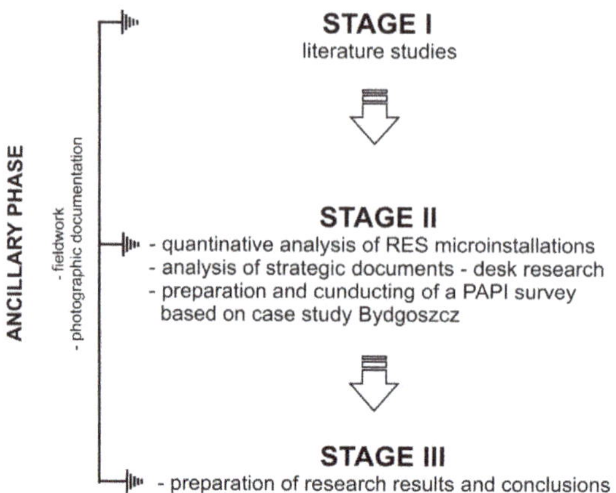


**Figure 1.** General research procedure framework. Source: own study.

In the next step in the research procedure, the authors used basic desk research based on existing (i.e., secondary) data [86]. It consisted in an analysis of the content of strategic documents (the Long-term National Development Strategy, the Strategy for Responsible Development [87], the Urban Development Strategy [88–107] and Low-Emission Economy Plans [108–126]) for the 20 largest cities in Poland, in terms of the smart city concept, with particular emphasis on provisions regarding the implementation of RES in the urban space. It was also investigated which of the analysed cities has a separate smart development strategy, and these documents were also examined. The authors based their analysis on documents in Polish available from the official websites of the cities selected for analysis. In total, 5480 records were analysed for entries related to RES in the context of the smart city.

The study's research period covered:

- the year 2019, for quantitative data relating to the installation of RES in cities;
- 2007–2020 for the analysis of strategic documents.

The article also uses the results of a survey that was conducted in Bydgoszcz (Kujawsko-Pomorskie voivodeship). The Bydgoszcz case study is a valuable testing ground for issues relating to a large post-communist (and post-industrial) city's energy transition towards becoming a smart city. The activities undertaken in Bydgoszcz constitute a starting point for addressing the challenges in effecting a smart city energy transition in cities marred by a past experience of being subject to a centrally planned economy. The scale of the selected case study is also crucial here, because, as contemporary research shows [78], energy transition is an extremely complex process, so research in this field should relate primarily to the local level, which in this case is the city. The authors used the PAPI survey method [127,128]. This method allows a questionnaire to include many research questions that are high in difficulty and complexity. The questionnaire form was based on closed questions using a dichotomous, nominal, and modified Likert scale [129]. In total, 475 questionnaires were collected, in accordance with the principle that the obtained sample should reflect the demographic and social structure of the city under study. The study results were digitized and analysed using IBM SPSS software. Another advantage is the high availability of respondents [130]. The study was carried out in the spring of 2019 in the city of Bydgoszcz, which was a case study and functioned as a research and analysis testing ground for the survey and a good example of the use of smart-city RES solutions. Bydgoszcz is also distinguished for its involvement in international initiatives relating to sustainable development and the dissemination of renewable energy sources. There is a RES Demonstration Centre in the city that teaches how individual installations work. Using Bydgoszcz as a case study allowed for an in-depth analysis of RES-based smart-city solutions and of how they are perceived by the local communities that benefit from "smart" changes [131].

## 3. Results

### 3.1. RES Installations in the Largest Cities of Poland

Taking into account the structure of RES facilities according to the number of installations in cities in Poland, it should be stated, the dominant energy source is solar energy (Figure 2). In the case of small installations, photovoltaic (PV) technologies account for almost 72% of all RES (Table 1). The undisputed leader in the use of solar technologies, including PV, is Silesia [132], with such cities as Katowice, Gliwice, and Zabrze. They top the national ranking in terms of number of PV installations. The average installed PV capacity does not exceed 0.5 MW. The growing popularity of solar technologies, including in the field of electricity production, is conditioned by several factors: they are usually located very close to both the producer and the consumer; they are the most environmentally friendly energy technologies; and they do not conflict with architectural and aesthetic solutions in buildings [133]. Moreover, they have the best public image of all RES [134].

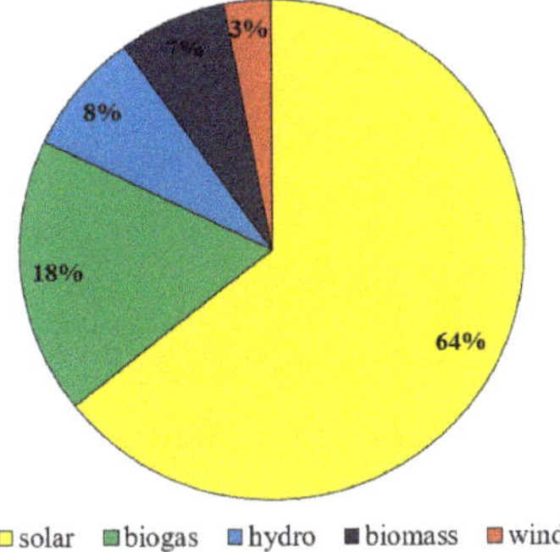


**Figure 2.** The structure of renewable energy installations according to the number of installations in major Polish cities. Source: own study based on data from the Energy Regulatory Office (ERO) [85].

Biogas also plays an important role in the structure of RES installations. Facilities of this type are based on processing waste—mainly in sewage treatment plants—and the average installed capacity is approximately 1 MW. Their advantages are in co-generation, meaning the production of both electricity and heat, and in allowing the principles of a circular economy to be implemented [135]. In third place was hydropower (Figure 3), whose development corresponds with the occurrence of appropriate environmental conditions. Half of the hydroelectric power plants located in the cities studied, including the largest (with a capacity of 5.5 MW), are to be found in Bydgoszcz. It is a city at a forking of rivers, including Poland's largest river, the Vistula, and it is criss-crossed by numerous canals.

Looking at the number of RES installations alone, it should be noted that the most (more than 10) are in Katowice (18), Warsaw (14), and Szczecin (12) (Figure 3). Szczecin also has the largest number of small renewable energy installations (11). The largest, though among the least numerous, are installations using biomass, including those that are co-fired by fossil fuels. Their installed capacity reaches, for example, 170 MW in Warsaw. It is in Warsaw that the highest total installed RES capacity was recorded, at 180.6 MW (Table 2). Warsaw follows the trends observed in many European cities, moving towards sustainable and smart development. Numerous activities are carried out in this city with the aim of improving the everyday functioning of the city, thus increasing the systematic share of RES in the overall energy balance.

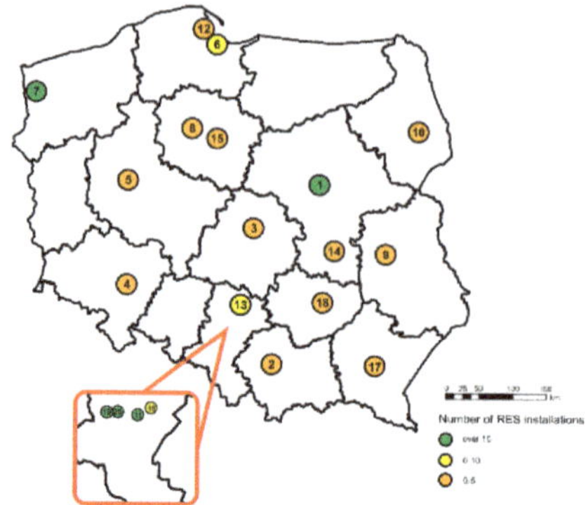

**Figure 3.** RES installations in cities in Poland. Explanations: 1 Warsaw, 2 Kraków, 3 Łódź, 4 Wrocław, 5 Poznań, 6 Gdańsk, 7 Szczecin, 8 Bydgoszcz, 9 Lublin, 10 Białystok, 11 Katowice, 12 Gdynia, 13 Częstochowa, 14 Radom, 15 Toruń, 16 Częstochowa, 17 Rzeszów, 18 Kielce, 19 Gliwice, 20 Zabrze. Source: own study based on data from ERO (2020) [85].

**Table 2.** RES installations in cities in Poland.

| City | Population | RES Installations | | | | | | | | |
|---|---|---|---|---|---|---|---|---|---|---|
| | | Number of RES Installations | Total Installed RES Capacity (MW) | Average Power of RES Installations (MW) | Small RES Installations * | | | | | |
| | | | | | Number of Installations by RES | | | | | |
| | | | | | Sum | Biogas | Biomass | Sun | Water | Wind |
| Warsaw | 1,790,658 | 14 | 180.60 | 12.90 | 4 | 0 | 0 | 4 | 0 | 0 |
| Kraków | 779,115 | 5 | 19.85 | 3.97 | 3 | 0 | 0 | 3 | 0 | 0 |
| Łódź | 679,941 | 4 | 59.36 | 14.84 | 5 | 0 | 0 | 5 | 0 | 0 |
| Wrocław | 642,869 | 1 | 0.07 | 0.07 | 2 | 0 | 0 | 0 | 2 | 0 |
| Poznań | 534,813 | 2 | 2.13 | 1.07 | 2 | 0 | 0 | 2 | 0 | 0 |
| Gdańsk | 470,907 | 7 | 4.97 | 0.71 | 4 | 0 | 0 | 3 | 0 | 1 |
| Szczecin | 401,907 | 12 | 93.47 | 7.79 | 11 | 2 | 0 | 7 | 2 | 0 |
| Bydgoszcz | 348,190 | 5 | 6.51 | 1.30 | 3 | 1 | 0 | 0 | 2 | 0 |
| Lublin | 339,784 | 3 | 1.73 | 0,58 | 1 | 0 | 0 | 1 | 0 | 0 |
| Białystok | 297,554 | | no data | | 2 | 0 | 0 | 2 | 0 | 0 |
| Katowice | 292,774 | 18 | 1.15 | 0.06 | 6 | 0 | 0 | 6 | 0 | 0 |
| Gdynia | 246,348 | 3 | 0.07 | 0.02 | 0 | 0 | 0 | 0 | 0 | 0 |
| Częstochowa | 220,433 | 9 | 2.21 | 0.25 | 4 | 0 | 0 | 3 | 1 | 0 |
| Radom | 211,371 | 4 | 0.95 | 0.24 | 1 | 0 | 0 | 0 | 0 | 1 |
| Toruń | 201,447 | 1 | 0.93 | 0.93 | 1 | 0 | 0 | 1 | 0 | 0 |
| Sosnowiec | 199,974 | 7 | 1.75 | 0.25 | 2 | 0 | 0 | 2 | 0 | 0 |
| Rzeszów | 196,208 | | no data | | 2 | 0 | 0 | 2 | 0 | 0 |
| Kielce | 194,852 | 2 | 6.73 | 3.36 | 0 | 0 | 0 | 0 | 0 | 0 |
| Gliwice | 178,603 | 12 | 1.35 | 0.11 | 6 | 1 | 0 | 3 | 2 | 0 |
| Zabrze | 172,360 | 11 | 78.33 | 7.12 | 5 | 2 | 0 | 3 | 0 | 0 |
| Sum | | 120 | 462.15 | 3.85 | 60 | 6 | 0 | 43 | 9 | 2 |

* The small RES Installations according to the Art. 8 s. 1 of the Act on renewable energy sources (Journal of Laws of 2018, item 1269, as amended) [84]. Source: own study based on data from ERO (2020) [85].

It is worth emphasizing that the number of power plants does not correspond with total installed capacity. Some renewable energy sources are typified by large-scale installations. These include biomass (which is often also co-fired by conventional raw materials) and biogas produced in municipal wastewater treatment plants. Of all RES installations, it is biogas plants that ensure the most stable, predictable, and efficient energy production. Their operation is not subject to fluctuations in natural conditions—as, for example, wind, solar or water power plants are—but depends primarily on human labor [80]. These large renewable energy technologies (which use biogas, but also biomass) result from the modernizing of existing key municipal facilities, including: municipal heat and power plants, wastewater treatment plants, and waste treatment sites. In turn, new RES investments include small-scale installations—mainly PV. One example is Katowice, which has the most RES installations of the cities studied, with a total capacity of 1.15 MW. Nevertheless, it is the development and growing popularity of place-based DG systems based on local resources that will determine the future of smart cities, ensuring the development of prosumer attitudes among city residents on the energy services market.

### 3.2. RES in City Development Documents

The development strategies of Polish cities increasingly refer to the smart city concept, as a result of national activities for smart urban development. The strategic goals of Polish policy documents, i.e., the Long-term National Development Strategy and the Strategy for Responsible Development, are: economic competitiveness and innovation; achieving sustainable development potential; sustained economic growth based increasingly on knowledge, data, and organizational excellence; and socially sensitive and territorially balanced development [87,136]. These top-down guidelines mobilize city authorities to implement intelligent solutions in urban spaces, including in the field of renewable energy sources.

The analyzed cities in Poland contain renewable energy provision in almost all their strategic documents, such as Urban Development Strategies and Low-Emission Economy Plans (Figure 4, Table S1 in Supplementary Materials). These two documents are of key importance in creating modern and ecological urban development in line with smart city principles. They usually concern improving energy efficiency and energy security—as is the case in Gdańsk, Kraków, Kielce, and Gliwice. An equally important goal chosen by Polish cities is to increase diversification of energy sources and to increase the use of energy from renewable sources (Toruń, Radom, Rzeszów). Another important goal in city strategies is to promote and disseminate RES and other energy-efficient solutions—as exemplified by Wrocław, Warsaw, Gdynia, and Sosnowiec. In turn, cities such as Poznań and Białystok have already indicated specific activities aimed at introducing modern, energy-saving technologies and solutions in public spaces and buildings, including using intelligent solutions for greater use of renewable energy.

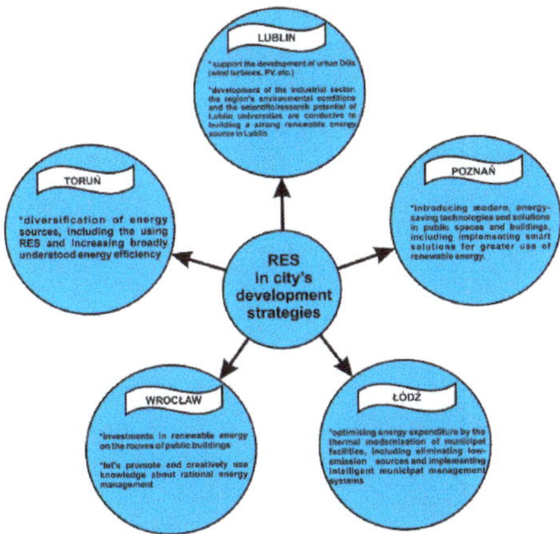


**Figure 4.** RES in urban development strategy. Source: own study based on urban development strategies [88–107].

In implementing RES in cities, low-emission economy plans are an important document, setting the path for the development of RES installations (Figure 5, Table S1 in Supplementary Materials). These documents already contain specific directions for renewable energy investments. Taking the example of Warsaw, new or adapted power plants generating electricity and heat in high-efficiency co-generation with RES are planned, as are new or adapted intelligent medium- and low-voltage distribution networks dedicated to increasing RES generation. It should be noted that most cities are focusing on solar energy, which is why many low-emission economy plans mention increasing the share of renewable energy in the total energy consumption by installing PV in public buildings (Rzeszów, Bydgoszcz, and Radom). In Kraków too, it is planned to install PVs, this time on bus roofs, and in Gdynia, on parking shelters at the trolleybus depot. In Łódź, however, it is planned to construct a reinforced-concrete passive office building and to install photovoltaic panels.

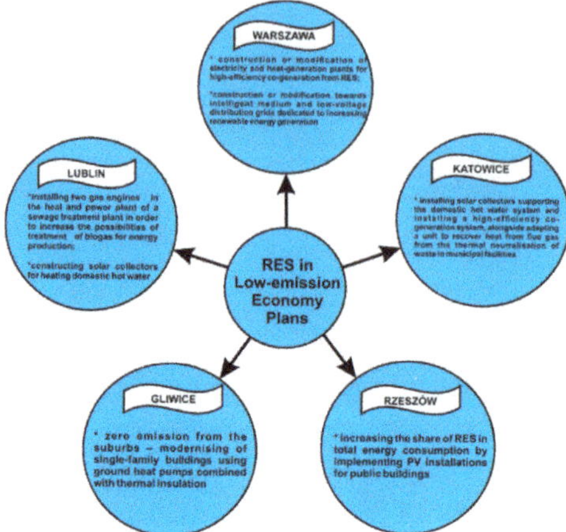


**Figure 5.** RES in low-emission economy plans. Source: own study based on low-emission economy plans [108–126].

It is worth emphasizing that some large cities in Poland also have separate documents outlining how to achieve the level of "smart city". Of the cities analyzed in this article, five have such documents: Warsaw, Kraków, Wrocław, Poznań, and Kielce. Smart city strategies also contain references to the desire to implement intelligent RES solutions (Figure 6). These are generally broad statements, such as about striving to increase the extent of RES use. However, in Kraków and Warsaw, for example, specific investments using PV are indicated.

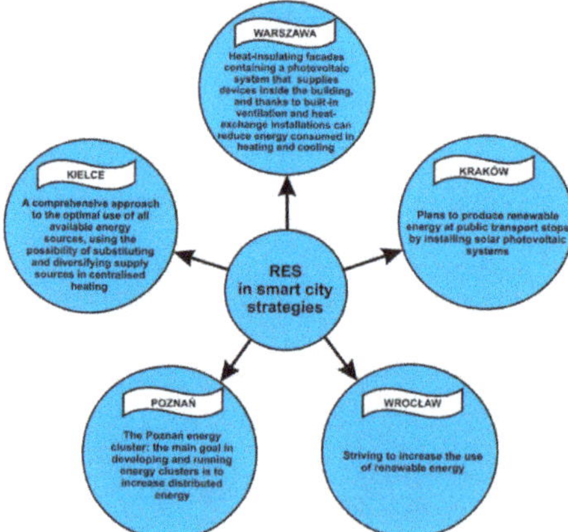


**Figure 6.** RSE in smart city strategies. Source: own study based on smart city strategies [137–141].

The conducted research shows that Polish cities focus first on safe and reliable energy supplies, and then on diversification of energy sources. Large global metropolises are applying a similar energy strategy [11,142–145]. It is much easier for city authorities to include only broad statements about increasing the share of renewable energy sources in strategic documents than to indicate specific types of investments in renewable energy sources and timeframes for their implementation.

### 3.3. Case Study: Bydgoszcz

Bydgoszcz was selected for detailed research as one of the greenest cities in Poland every year. For instance, in 2019, Bydgoszcz took second place in terms of the share of green areas in the total area among the studied Polish cities [146]. It is also distinguished for its involvement in international initiatives relating to sustainable development and the dissemination of renewable energy sources. There is also a RES Demonstration Centre in the city that teaches how individual installations work. Bydgoszcz is a member of the Association of Municipalities Polish Network "Energie Cités" [147], an international organization working to adapt cities to climate change, including supporting the development of distributed generation (a DG system). It has received distinctions in international competitions: in 2020, the city won the international Eco-City 2020 competition in the energy efficiency category. The implementation of RES projects in diverse facilities within the city was singled out for praise [148]. There are a total of six on-grid RES installations in Bydgoszcz, i.e., those selling generated energy to the national power grid. These are hydroelectric plants and a biogas plant [85].

This raises the question of what other renewable energy installations exist within the city and what attitude Bydgoszcz residents have towards new energy solutions. A total of 18.4% of respondents claim to use energy generated from RES, mainly in their homes. This is mainly solar energy (54.2%) in the form of solar collectors for the production of heat and domestic hot water. This small share of people claiming to use renewable energy technologies is due to the respondents (and city residents in general) mainly living in apartments in multi-family buildings (70% of respondents). Inhabitants of multi-family buildings have limited options for individual investment in renewable energy. Such decisions are shifted from the individual level to the level of housing cooperatives or communities, i.e., to building administrators.

This situation does not mean that RES installations are absent from buildings and public spaces in Bydgoszcz (Figure 7). On the contrary, in the urban space, there are various locations with RES installations, as confirmed by half of the respondents. The respondents mainly mentioned small and micro RES installations in autonomous power systems (Figure 8). Solar-powered road signs or hybrid devices (solar and wind energy) dominate. These installations are used to improve the visibility and legibility of existing road signs, and to increase delivery of information by installing additional signals and devices. Illuminated road signs are located in particularly sensitive places, such as pedestrian crossings, junctions, and bends in the road. In addition, the respondents noted renewable energy installations supplying city bike stations and lighting for bus and tram stops. The use of off-grid RES devices also enables the operation of vehicles and means of transport, such as the Bydgoszcz Water Tram that runs as part of the municipal public transport system and is a local tourist attraction.

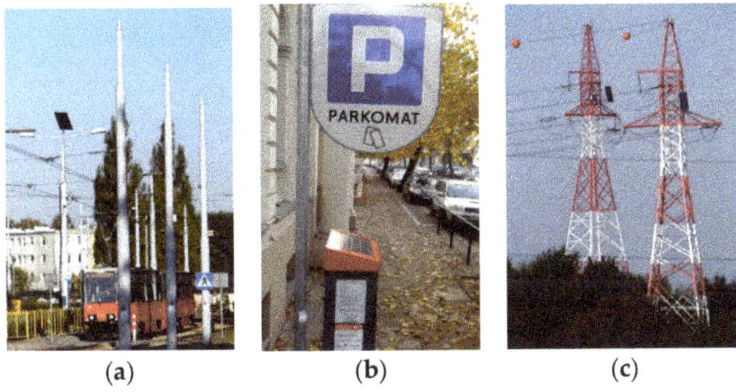


(a) (b) (c)

**Figure 7.** Examples of small PV autonomous installations. (**a**) Tramway infrastructural feature, Bydgoszcz, Glinki district. (**b**) Parking meter, Bydgoszcz, Śródmieście district. (**c**) Aeronautical infrastructure feature, Bydgoszcz, Wzgórze Wolności district. Source: own study.

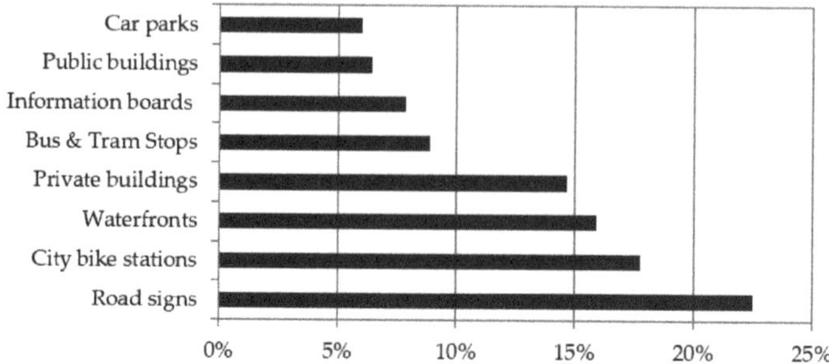


**Figure 8.** Breakdown of answers to the question "Have you seen renewable energy installations within the city of Bydgoszcz?" Source: own study.

According to the respondents, it is inhabitants who should have the greatest influence on the direction that RES development takes in the city. Thus, when asked as potential decision-makers what types of RES have the greatest growth potential in Bydgoszcz, they mention hydro energy in first place (Figure 9). Respondents' arguments for developing hydropower in Bydgoszcz emphasize the city's waterside location (its rivers and numerous canals) and the existing and prospering hydro-electric power plants, including two small facilities. It is also important to focus the municipal policy in Bydgoszcz on revitalizing waterfronts and emphasizing their priority role in the urban fabric. They also see the potential of solar energy as an inexhaustible and environmentally friendly source of energy, whose installations can be placed on building roofs and facades.

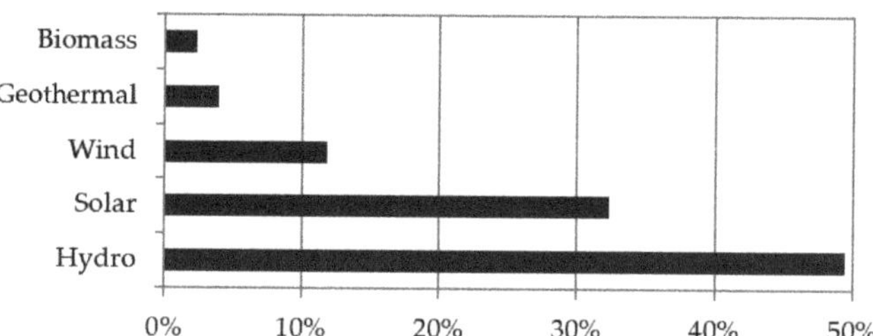


**Figure 9.** Breakdown of answers to the question "Which renewable energy sources, in your opinion, have the best growth potential in Bydgoszcz?" Source: own study.

Renewable energy installations are an increasingly common feature of the urban landscape of Bydgoszcz. They are developing mainly in the public sphere of the city, where the use of solar energy dominates in the form of off-grid systems. Solar energy (solar thermal collectors and photovoltaics) are also popular in private and public buildings to meet consumer energy needs. Due to its natural conditions, on-grid hydropower is also being developed in Bydgoszcz.

## 4. Discussion

Progressive urbanization is pushing the search for modern solutions to maintain balance in urban ecosystems. As mentioned in the introduction, demographic growth and the increased activity of various economic sectors in cities are increasing the demand for electricity. Today, there are sustainable urban development strategies aimed at maintaining harmony between the economy, society, and the environment. Rapid changes in these three areas, as well as the need for cities to adapt to climate change, are requiring that cities transform. Cities are unlikely to be transformed in absolute accordance with a single concept of urban development such as the concept of sustainable development, of the digital city, of the eco-city, of the low-emission city, or of the smart city. Rather, all these concepts are affecting cities, and will also do so in the future.

Smart cities also need to develop in a sustainable manner, and thus be environmentally friendly by reducing harmful emissions and switching to renewable energy sources [149,150]. Hence, providing cities with safe and permanent access to energy seems to be the issue of key importance. As Kammen and Sunter (2016) [151] noted in their research, a particular challenge is posed by the limited access to areas where energy installations may be located in urban areas. The balance between the high energy demand in cities and the energy density provided by renewable sources should therefore be the starting point for designing any analytical framework for a decarbonized urban space. Moreover, many studies show that cities' potential for RES generation remains untapped [152–155] This is the case in cities in the UK, for example, where the potential of solar and wind energy is not being exploited [156]. The situation in Polish cities is similar.

The presented results are evidence of the spread of RES in Polish cities. This is also confirmed by other Polish RES research [25,82,157]. There are similar trends in the popularization of urban RES in Poland's neighbor, Germany. There, it is being popularized via the energy transition known as *Energiewende*, which refers to the ongoing energy experiment to create sustainable energy transitions (SETs) by (radically and) increasingly selecting renewable energy sources and systems while abandoning the unsustainable use of energy resources [158,159]. This is also evident in German cities, the best example of which is Munich.

It should also be emphasized that the popularity of photovoltaic installations is increasing in Polish cities, which is a very positive sign in the context of a smart city. These installations are very cost effective and are having a real effect on municipal budgets. The economic aspect of PV has been emphasized by, for example, Abrao et al. (2017) [160]. The significant increase in the number of solar PV roof installations has made buildings the largest source of urban space available for deployment [161]. It should also be noted that the growing threats of climate change and the global challenges of sustainable development, in combination with the significant decrease in the cost of renewable energy sources, has led to solar power systems being recognized as a major feature of a mitigation strategy [152].

At this point, it is worth emphasizing that some of the analysed cities in Poland have their own smart development strategies that mention the development of RES, considering it to be a priority in becoming a smart city. This approach is not surprising as, by the end of 2018, more than 230 cities around the world had adopted targets for 100% renewable energy in at least one sector of the economy [162]. Therefore, the present study confirms the energy transformation that is being seen to be taking place in many cities around the world in the spirit of sustainable, smart development. Thus, the future success of RES development in Polish cities will be determined by properly conducted urban policies that focus on smart solutions and on raising residents' awareness of "green energy", which in addition to its economic benefits, carries the recently popular environmentally friendly message of saving a degraded planet.

## 5. Conclusions

Moving from conventional to renewable energy is an extremely complex and problematic process, both conceptually (creating strategies and action plans) and in practice (developing RES installations). Additionally, in Poland, the pace and scope of the energy transition are hampered by historical factors related to the centrally planned socialist economy. Nevertheless, in cities in Poland, the local authorities and the local community are interested in and disposed towards renewable energy sources. All the development strategies and low-emission economy plans of the analyzed cities contain provisions referring favourably to RES. However, there are some disproportions in this respect, because it is far easier to address RES in a general manner in strategic documents than to indicate specific investments planned in the urban space. The superficiality of assumptions relating to the growing importance of RES in urban strategies may cause concern. On the one hand, the imprecision of the municipal documents allows for ongoing changes to RES projects being implemented, which is especially relevant given that the energy transition has diverse implications: from spatial, political, and socio-economic aspects, to structural changes. Meanwhile, it may also offer some security in the face of changes to energy policy and to current legislative and fiscal instruments for developing renewable energy. Given that actions stimulating the energy transition should be integrated into the local context and programmed with regard to local conditions, far more clearly formulated place-based planning assumptions are needed.

In addressing the article's titular question, it should be stated that renewable energy installations are a reality in cities in Poland and they are increasingly becoming important features within urban spaces, as evidenced by the conducted analyses. These changes vary in pace and scope, as do the conditions in which cities function. We see a progressive diversification of energy sources. Nevertheless, the most popular is photovoltaics, which is being used in infrastructure dedicated both to buildings (public and private) and to autonomous power systems. Small-scale RES installations are spreading fast. Meanwhile, large-scale ones are the result of the modernization of municipal sewage treatment plants, municipal heat and power plants, and waste processing sites, where biogas and biomass are used in energy production. The identified place-based DG facilities prove that the smart city idea is increasing in popularity in Polish cities. This situation should be seen as an opportunity for cities, including post-socialist ones, to create a modern, functional, and environmentally friendly city (a smart city), and thus to build international market competitiveness. It must be noted that the concept of smart city should be a pillar of further development of Polish cities—it is the present reality, the reality of the future.

**Supplementary Materials:** The following are available online at http://www.mdpi.com/1996-1073/13/21/5795/s1, Table S1: RES in city development documents.

**Author Contributions:** Conceptualization, A.L. and J.C.-M.; methodology, A.L., J.C.-M. and K.R.; formal analysis, A.L., J.C.-M. and K.R.; data curation, J.C.-M. and T.S.; writing—original draft preparation, A.L., J.C.-M., K.R. and T.S.; writing—review and editing, A.L., J.C.-M., K.R. and T.S.; visualization, J.C.-M. and T.S. All authors have read and agreed to the published version of the manuscript.

**Funding:** This research was funded by National Science Centre, Poland, Project no. 2015/19/N/HS4/02586: Ecologization of cities in Poland in the light of selected parameters of sustainable development and 2016/21/D/HS4/00714: Biogas enterprises from the perspective of the embeddedness concept.

**Acknowledgments:** We would like to express our gratitude to support University Center of Excellence "Interacting Minds, Societies, Environment (IMSErt) and the respondents of our survey for supporting data necessary for this research.

**Conflicts of Interest:** The authors declare no conflict of interest.

## References

1. Caragliu, A.; del Bo, C.; Nijkamp, P. Smart cities in Europe. *J. Urban Technol.* **2011**, *18*, 65–82. [CrossRef]
2. Albino, V.; Berardi, U.; Dangelico, R.M. Smart cities: Definitions, dimensions, performance, and initiatives. *J. Urban Technol.* **2015**, *22*, 3–21. [CrossRef]
3. Ahvenniemi, H.; Huovila, A.; Pinto-Seppä, I.; Airaksinen, M. What are the differences between sustainable and smart cities? *Cities* **2017**, *60*, 234–245. [CrossRef]
4. Douša, M.; Lewandowska, A. Inteligentní města v praxi: Projekty realizované v slovenských a polských městech. In *Trvalo Udržate'ný Rozvoj v Krajinách Európskej Únie, Nekonferenčný Zborník Vedeckých Prác: VEGA č. 1/0302/18: Inteligentné Mestá ako Spôsob Implementácie Konceptu Trvalo Udržate'ného Rozvoja Miest SR*; Čepelová, A., Koreňová, D., Eds.; Univerzita Pavla Jozefa Šafárika: Košice, Slovak, 2019; pp. 171–189.
5. Kumar, H.; Singh, M.K.; Gupta, M.P.; Madaan, J. Moving towards smart cities: Solutions that lead to the smart city transformation framework. *Technol. Forecast. Soc. Chang.* **2020**, *153*, 1–16. [CrossRef]
6. Leźnicki, M.; Lewandowska, A. Contemporary concepts of a city in the context of sustainable development: Perspective of humanities and natural sciences. *Probl. Ekorozw.* **2016**, *11*, 45–54.
7. Kumar, T.V.; Dahiya, B. Smart economy in smart cities. In *Smart Economy in Smart Cities*; Springer: Berlin/Heidelberg, Germany, 2017; pp. 3–76.
8. Visvizi, A.; Lytras, M.D. Rescaling and refocusing smart cities research: From mega cities to smart villages. *J. Sci. Technol. Policy Manag.* **2018**, *9*, 134–145. [CrossRef]

9. Ruhlandt, R.W.S. The governance of smart cities: A systematic literature review. *Cities* **2018**, *81*, 1–23. [CrossRef]
10. Cardullo, P.; Kitchin, R. Smart urbanism and smart citizenship: The neoliberal logic of 'citizen-focused'smart cities in Europe. Environment and Planning. *Politics Space* **2019**, *37*, 813–830. [CrossRef]
11. Strielkowski, W.; Veinbender, T.; Tvaronavičienė, M.; Lace, N. Economic efficiency and energy security of smart cities. *Econ. Res. Ekon. Istraživanj* **2020**, *33*, 788–803. [CrossRef]
12. Janik, A.; Ryszko, A.; Szafraniec, M. Scientific landscape of smart and sustainable cities literature: A bibliometric analysis. *Sustainability* **2020**, *12*, 779. [CrossRef]
13. Nam, T.; Pardo, T.A. Conceptualizing smart city with dimensions of technology, people, and institutions. In Proceedings of the 12th Annual International Digital Government Research Conference: Digital Government Innovation in Challenging Times, College Park, MD, USA, 12–15 June 2011; pp. 282–291.
14. Lee, J.H.; Phaal, R.; Lee, S.H. An integrated service-device-technology roadmap for smart city development. *Technol. Forecast. Soc. Chang.* **2013**, *80*, 286–306. [CrossRef]
15. Akaraci, S.; Usman, M.A.; Usman, M.R.; Ahn, D.J. From smart to smarter cities: Bridging the dimensions of technology and urban planning. In Proceedings of the 2016 International Conference on Smart Green Technology in Electrical and Information Systems (ICSGTEIS), Bali, Indonesia, 6–8 October 2016; pp. 74–78.
16. Sepasgozar, S.M.; Hawken, S.; Sargolzaei, S.; Foroozanfa, M. Implementing citizen centric technology in developing smart cities: A model for predicting the acceptance of urban technologies. *Technol. Forecast. Soc. Chang.* **2019**, *142*, 105–116. [CrossRef]
17. Shapiro, J.M. Smart cities: Quality of life, productivity, and the growth effects of human capital. *Rev. Econ. Stat.* **2016**, *88*, 324–335. [CrossRef]
18. Jurenka, R.; Cagáňová, D.; Horňáková, N.; Stareček, A. Smart city in terms of social innovations and human capital. *Smart City* **2016**, *360*, 2nd. [CrossRef]
19. Giffinger, R.; Fertner, C.; Kramar, H.; Kramar, H.; Kalasek, R.; Pichler-Milanovic, N.; Meijers, E. Smart Cities. In *Ranking of European Medium-Sized Cities*; Centre for Regional Science, Vienna University of Technology: Viena, Austria, 2017; Available online: http://www.smart-cities.eu/download/smart_cities_final_report.pdf (accessed on 12 September 2020).
20. Jonek-Kowalska, I.; Kaźmierczak, J.; Kramarz, M.; Hilarowicz, A.; Wolny, M. Introduction to the research project "Smart City: A holistic approach". In Proceedings of the 5th SGEM International Multidisciplinary Scientific Conferences on SOCIAL SCIENCES and ARTS SGEM 2018, Sofia, Bulgaria, 26 August–1 September 2018; pp. 101–112. [CrossRef]
21. Kusch-Brandt, S. Urban Renewable Energy on the Upswing: A Spotlight on Renewable Energy in Cities in REN21's "Renewables 2019 Global Status Report". *Resources* **2019**, *8*, 139. [CrossRef]
22. Maier, S.; Narodoslawsky, M. Optimal Renewable Energy Systems for Smart Cities. In Proceedings of the 24th European Symposium on Computer Aided Process Engineering—ESCAPE 24, Budapest, Hungary, 15–18 June 2014; Klemes, J., Ed.; University of Pannonia: Veszprem, Hungary, 2014; pp. 1849–1854.
23. Lewandowska, A.; Chodkowska-Miszczuk, J.; Rogatka, K. Development of renewable energy sources in big cities in Poland in the context of urban policy. *Renew. Energy Sources Eng. Technol. Innov.* **2019**, 842–852. [CrossRef]
24. Eicker, U. Renewable Energy Sources within Urban Areas: Results from European Case Studies. *Ashrae Trans.* **2012**, *118*, 73–80.
25. Gerpott, T.J.; Paukert, M. Determinants of willingness to pay for smart meters: An empirical analysis of household customers in Germany. *Energy Policy* **2013**, *61*, 483–495. [CrossRef]
26. Gabillet, P. Energy supply and urban planning projects: Analysing tensions around district heating provision in a French eco-district. *Energy Policy* **2015**, *78*, 189–197. [CrossRef]
27. Kılkış, Ş. Sustainable development of energy, water and environment systems index for Southeast European cities. *J. Clean. Prod.* **2016**, *130*, 222–234. [CrossRef]
28. Petersen, J.P. Energy concepts for self-supplying communities based on local and renewable energy sources: A case study from northern Germany. *Sustain. Cities Soc.* **2016**, *26*, 1–8. [CrossRef]
29. Kazak, J.; Van Hoof, J.; Szewranski, S. Challenges in the wind turbines location process in Central Europe–The use of spatial decision support systems. *Renew. Sustain. Energy Rev.* **2017**, *76*, 425–433. [CrossRef]

30. Ahas, R.; Mooses, V.; Kamenjuk, P.; Tamm, R. Retrofitting Soviet-Era Apartment Buildings with 'Smart City'Features: The H2020 SmartEnCity Project in Tartu, Estonia. In *Housing Estates in the Baltic Countries*; Springer Science and Business Media LLC: Berlin, Germany, 2019; p. 357.
31. Bahers, J.B.; Tanguy, A.; Pincetl, S. Metabolic relationships between cities and hinterland: A political-industrial ecology of energy metabolism of Saint-Nazaire metropolitan and port area (France). *Ecol. Econ.* **2020**, *167*, 106447. [CrossRef]
32. Hammer, S.A. Renewable energy policymaking in New York and London: Lessons for other 'World Cities'? In *Urban Energy Transition*; Elsevier: Amsterdam, The Netherlands, 2008; pp. 141–172.
33. Denis, G.S.; Parker, P. Community energy planning in Canada: The role of renewable energy. *Renew. Sustain. Energy Rev.* **2009**, *13*, 2088–2095. [CrossRef]
34. Moscovici, D.; Dilworth, R.; Mead, J.; Zhao, S. Can sustainability plans make sustainable cities? The ecological footprint implications of renewable energy within Philadelphia's Greenworks Plan. *Sustain. Sci. Pract. Policy* **2015**, *11*, 32–43.
35. Bagheri, M.; Shirzadi, N.; Bazdar, E.; Kennedy, C.A. Optimal planning of hybrid renewable energy infrastructure for urban sustainability: Green Vancouver. *Renew. Sustain. Energy Rev.* **2018**, *95*, 254–264.
36. DeRolph, C.R.; McManamay, R.A.; Morton, A.M.; Nair, S.S. City energy sheds and renewable energy in the United States. *Nat. Sustain.* **2019**, *2*, 412–420.
37. Hess, D.J.; Gentry, H. 100% renewable energy policies in US cities: Strategies, recommendations, and implementation challenges. *Sustain. Sci. Pract. Policy* **2019**, *15*, 45–61.
38. Kouhestani, F.M.; Byrne, J.; Johnson, D.; Spencer, L.; Hazendonk, P.; Brown, B. Evaluating solar energy technical and economic potential on rooftops in an urban setting: The city of Lethbridge, Canada. *Int. J. Energy Environ. Eng.* **2019**, *10*, 13–32.
39. Ramírez, A.M.; Sebastian, P.J.; Gamboa, S.A.; Rivera, M.A.; Cuevas, O.; Campos, J. A documented analysis of renewable energy related research and development in Mexico. *Int. J. Hydrog. Energy* **2000**, *25*, 267–271.
40. Huacuz, J.M. The road to green power in Mexico—reflections on the prospects for the large-scale and sustainable implementation of renewable energy. *Energy Policy* **2005**, *33*, 2087–2099. [CrossRef]
41. de Araújo, M.S.M.; de Freitas, M.A.V. Acceptance of renewable energy innovation in Brazil—Case study of wind energy. *Renew. Sustain. Energy Rev.* **2008**, *12*, 584–591. [CrossRef]
42. Fonseca, J.A.; Schlueter, A. Novel approach for decentralized energy supply and energy storage of tall buildings in Latin America based on renewable energy sources: Case study–Informal vertical community Torre David, Caracas–Venezuela. *Energy* **2013**, *53*, 93–105. [CrossRef]
43. Cedeno, M.L.D.; Arteaga, M.G.D.; Perez, A.V.; Arteaga, M.L.D. Regulatory framework for renewable energy sources in Ecuador case study province of Manabi. *Int. J. Soc. Sci. Humanit.* **2017**, *1*, 29–42. [CrossRef]
44. Pérez-Denicia, E.; Fernández-Luqueño, F.; Vilariño-Ayala, D.; Montaño-Zetina, L.M.; Maldonado-López, L.A. Renewable energy sources for electricity generation in Mexico: A review. *Renew. Sustain. Energy Rev.* **2017**, *78*, 597–613. [CrossRef]
45. Lino, F.A.M.; Ismail, K.A.R. Evaluation of the treatment of municipal solid waste as renewable energy resource in Campinas, Brazil. *Sustain. Energy Technol. Assess.* **2018**, *29*, 19–25. [CrossRef]
46. Bugaje, I.M. Renewable energy for sustainable development in Africa: A review. *Renew. Sustain. Energy Rev.* **2006**, *10*, 603–612. [CrossRef]
47. Cloutier, M.; Rowley, P. The feasibility of renewable energy sources for pumping clean water in sub-Saharan Africa: A case study for Central Nigeria. *Renew. Energy* **2011**, *36*, 2220–2226. [CrossRef]
48. Zawilska, E.; Brooks, M.J. An assessment of the solar resource for Durban, South Africa. *Renew. Energy* **2011**, *36*, 3433–3438. [CrossRef]
49. Gumbo, T. Scaling up sustainable renewable energy generation from municipal solid waste in the African continent: Lessons from EThekwini, South Africa. *Consilience* **2014**, *12*, 46–62.
50. Akuru, U.B.; Onukwube, I.E.; Okoro, O.I.; Obe, E.S. Towards 100% renewable energy in Nigeria. *Renew. Sustain. Energy Rev.* **2017**, *71*, 943–953. [CrossRef]
51. Bouhal, T.; Agrouaz, Y.; Kousksou, T.; Allouhi, A.; El Rhafiki, T.; Jamil, A.; Bakkas, M. Technical feasibility of a sustainable Concentrated Solar Power in Morocco through an energy analysis. *Renew. Sustain. Energy Rev.* **2018**, *81*, 1087–1095. [CrossRef]
52. Jebaraj, S.; Iniyan, S. Renewable energy programmes in India. *Int. J. Glob. Energy Issues* **2006**, *26*, 232–257. [CrossRef]

53. Bilgen, S.; Keleş, S.; Kaygusuz, A.; Sarı, A.; Kaygusuz, K. Global warming and renewable energy sources for sustainable development: A case study in Turkey. *Renew. Sustain. Energy Rev.* **2008**, *12*, 372–396. [CrossRef]
54. Cheng, H.; Hu, Y. Municipal solid waste (MSW) as a renewable source of energy: Current and future practices in China. *Bioresour. Technol.* **2010**, *101*, 3816–3824. [CrossRef]
55. Farooq, M.K.; Kumar, S. An assessment of renewable energy potential for electricity generation in Pakistan. *Renew. Sustain. Energy Rev.* **2013**, *20*, 240–254. [CrossRef]
56. Schroeder, P.M.; Chapman, R.B. Renewable energy leapfrogging in China's urban development? Current status and outlook. *Sustain. Cities Soc.* **2014**, *11*, 31–39. [CrossRef]
57. Madakam, S.; Ramaswamy, R. Sustainable smart city: Masdar (UAE) (A city: Ecologically balanced). *Indian J. Sci. Technol.* **2016**, *9*, 1–8. [CrossRef]
58. Noorollahi, Y.; Itoi, R.; Yousefi, H.; Mohammadi, M.; Farhadi, A. Modeling for diversifying electricity supply by maximizing renewable energy use in Ebino city southern Japan. *Sustain. Cities Soc.* **2017**, *34*, 371–384. [CrossRef]
59. Yuan, X.C.; Lyu, Y.J.; Wang, B.; Liu, Q.H.; Wu, Q. China's energy transition strategy at the city level: The role of renewable energy. *J. Clean. Prod.* **2018**, *205*, 980–986. [CrossRef]
60. Awan, A.B. Performance analysis and optimization of a hybrid renewable energy system for sustainable NEOM city in Saudi Arabia. *J. Renew. Sustain. Energy* **2019**, *11*, 025905. [CrossRef]
61. Fraser, T. Japan's resilient, renewable cities: How socioeconomics and local policy drive Japan's renewable energy transition. *Environ. Politics* **2019**, 500–523. [CrossRef]
62. Meng, N.; Xu, Y.; Huang, G. A stochastic multi-objective optimization model for renewable energy structure adjustment management–A case study for the city of Dalian, China. *Ecol. Indic.* **2019**, *97*, 476–485. [CrossRef]
63. Mithraratne, N. Roof-top wind turbines for microgeneration in urban houses in New Zealand. *Energy Build.* **2009**, *41*, 1013–1018. [CrossRef]
64. Martin, N.J.; Rice, J.L. Developing renewable energy supply in Queensland, Australia: A study of the barriers, targets, policies and actions. *Renew. Energy* **2012**, *44*, 119–127. [CrossRef]
65. White, L.V.; Lloyd, B.; Wakes, S.J. Are Feed-in Tariffs suitable for promoting solar PV in New Zealand cities? *Energy Policy* **2013**, *60*, 167–178. [CrossRef]
66. Dowling, R.; McGuirk, P.; Bulkeley, H. Retrofitting cities: Local governance in Sydney, Australia. *Cities* **2014**, *38*, 18–24. [CrossRef]
67. Imteaz, M.A.; Ahsan, A. Solar panels: Real efficiencies, potential productions and payback periods for major Australian cities. *Sustain. Energy Technol. Assess.* **2018**, *25*, 119–125. [CrossRef]
68. Li, H.X.; Edwards, D.J.; Hosseini, M.R.; Costin, G.P. A review on renewable energy transition in Australia: An updated depiction. *J. Clean. Prod.* **2020**, *242*, 118475. [CrossRef]
69. Skiba, M.; Mrówczyńska, M.; Bazan-Krzywoszańska, A. Modeling the economic dependence between town development policy and increasing energy effectiveness with neural networks. Case study: The town of Zielona Góra. *Appl. Energy* **2017**, *188*, 356–366. [CrossRef]
70. Stoeglehner, G.; Niemetz, N.; Kettl, K.H. Spatial dimensions of sustainable energy systems: New visions for integrated spatial and energy planning. *EnergySustain. Soc.* **2011**, *1*, 1–9. [CrossRef]
71. Stoeglehner, G.; Neugebauer, G.; Erker, S.; Narodoslawsky, M. *Integrated Spatial and Energy Planning: Supporting Climate Protection and the Energy Turn with Means of Spatial Planning*; Springer: Berlin/Heidelberg, Germany, 2016.
72. De Pascali, P.; Bagaini, A. Energy Transition and Urban Planning for Local Development. A Critical Review of the Evolution of Integrated Spatial and Energy Planning. *Energies* **2019**, *12*, 35. [CrossRef]
73. Asarpota, K.; Nadin, V. Energy Strategies, the Urban Dimension, and Spatial Planning. *Energies* **2020**, *13*, 3642. [CrossRef]
74. Adil, A.M.; Ko, Y. Socio-technical evolution of Decentralized Energy Systems: A critical review and implications for urban planning and policy. *Renew. Sustain. Energy Rev.* **2016**, *57*, 1025–1037. [CrossRef]
75. Ziembicki, P.; Klimczak, M.; Bernasiński, J. Optimization of municipal energy systems with the use of an intelligent analytical system. *Civil Environ. Eng. Rep.* **2018**, *28*, 132–144. [CrossRef]
76. Neves, D.; Baptista, P.; Simoes, M.; Silva, C.A.; Figueira, J.R. Designing a municipal sustainable energy strategy using multi-criteria decision analysis. *J. Clean. Prod.* **2018**, *176*, 251–260. [CrossRef]
77. Ceglia, F.; Esposito, P.; Marrasso, E.; Sasso, M. From smart energy community to smart energy municipalities: Literature review, agendas and pathways. *J. Clean. Prod.* **2020**, *254*, 120118. [CrossRef]

78. Young, J.; Brans, M. Fostering a local energy transition in a post-socialist policy setting. *Environ. Innov. Soc. Transit.* **2020**, *36*, 221–235.
79. Neofytou, H.; Nikas, A.; Doukas, H. Sustainable energy transition readiness: A multicriteria assessment index. *Renew. Sustain. Energy Rev.* **2020**, *131*, 109988. [CrossRef]
80. Chodkowska-Miszczuk, J. *Przedsiębiorstwa Biogazowe w Rozwoju Lokalnym w Świetle Koncepcji Zakorzenienia;* Wydawnictwo Uniwersytetu Mikołaja Kopernika: Toruń, Poland, 2019.
81. Delponte, I.; Schenone, C. RES Implementation in Urban Areas: An Updated Overview. *Sustainability* **2020**, *12*, 382. [CrossRef]
82. Chodkowska-Miszczuk, J. Small-Scale Renewable Energy Systems in the Development of Distributed Generation in Poland. *Morav. Geogr. Rep.* **2014**, *22*, 34–43. [CrossRef]
83. Sanchez-Miralles, A.; Calvillo, C.; Martín, F.; Villar, J. Use of Renewable Energy Systems in Smart Cities. *Green Energy Technol.* **2014**, *2*, 341–370.
84. *Ustawa o Odnawialnych Źródła Energii Z*; 20 lutego 2015 r., Dz. U. 2015 poz. 478 z późn: Warszawa, Poland, 2015.
85. Rejestr Wytwórców Energii w Małej Instalacji (10.09.2020), Urząd Regulacji Energetyki. Available online: https://bip.ure.gov.pl/bip/rejestry-i-bazy/wytworcy-energii-w-male/2138,Rejestr-wytworcow-energii-w-malej-instalacji.html (accessed on 15 September 2020).
86. Makowska, M. *Analiza Danych Zastanych. Przewodnik dla Studentów*; Scholar: Warszawa, Poland, 2012.
87. Strategia na Rzecz Odpowiedzialnego Rozwoju do Roku 2020 (z Perspektywą do 2030 r.). Available online: https://www.miir.gov.pl/media/48672/SOR.pdf (accessed on 15 September 2020).
88. Strategia Rozwoju Miasta Kielce na Lata 2007–2020. Available online: http://www.um.kielce.pl/gfx/kielce2/userfiles/files/pliki/strategia-rozwoju-miasta-kielce-aktualizacja-15092016.pdf (accessed on 12 September 2020).
89. Strategia Rozwoju Miasta Sosnowca do 2020 r. Available online: http://www.sosnowiec.pl/_upload/strategia2020.pdf (accessed on 12 September 2020).
90. Strategia Rozwoju Miasta Radomia na Lata 2008–2020. Available online: http://www.radom.pl/data/other/strategia_rozwoju_miasta_radomia_na_lata.pdf (accessed on 12 September 2020).
91. Strategia Rozwoju Miasta Białegostoku. Available online: https://www.bialystok.pl/pl/dla_biznesu/rozwoj_miasta/ (accessed on 12 September 2020).
92. Szczecin dla Ciebie Strategia Rozwoju Szczecina 2025. Available online: http://bip.um.szczecin.pl/chapter_11124.asp?soid=8ED6AD35235F4C07B05B8D5F81CF4090 (accessed on 12 September 2020).
93. Strategia zintegrowanego rozwoju Łodzi 2020+. Available online: https://uml.lodz.pl/dla-mieszkancow/o-miescie/strategia-lodzi-i-planowanie/strategia-rozwoju-lodzi/ (accessed on 12 September 2020).
94. Strategia rozwoju Miasta Lublin. Available online: https://bip.lublin.eu/gfx/bip/userfiles/_public/import/urzad-miasta-lublin/ogloszenia/konsultacje-spoleczne/2013/konsultacje-spoleczne-z-mieszk/74685_strategia_rozwoju_lublina_na_lata_2013_2020.diagnoza_sta.pdf (accessed on 12 September 2020).
95. Strategia Zintegrowanego i Zrównoważonego Rozwoju Miasta Gliwice do Roku 2022. 2014. Available online: https://bip.gliwice.eu/pub/html/um/files/2014-03-20_Strategia_uchwalona%282%29.pdf (accessed on 12 September 2020).
96. Gdańsk 2030 Plus Strategia Rozwoju Miasta. Available online: https://gdansk.xgcmy.pl/s1/d/20150158301/1.pdf (accessed on 12 September 2020).
97. Strategia Rozwoju Miasta "Katowice 2030". Available online: https://bip.katowice.eu/UrzadMiasta/ZamierzeniaIProgramy/dokument.aspx?idr=96518&menu=633 (accessed on 12 September 2020).
98. Strategia Rozwoju Miasta Częstochowa 2030+. Available online: http://www.czestochowa.pl/page/4859,strategia-rozwoju-miasta-2030-+-.html (accessed on 12 September 2020).
99. Strategia Rozwoju Rzeszowa do 2025. Available online: https://bip.erzeszow.pl/ (accessed on 12 September 2020).
100. Strategia Rozwoju Miasta Poznania 2020+. Available online: https://bip.poznan.pl/bip/strategia-rozwoju-miasta-poznania-2020,doc,42/strategia-rozwoju-miasta-poznania-2020,80837.html (accessed on 12 September 2020).
101. Strategia Rozwoju Miasta Gdyni 2030. Available online: http://2030.gdynia.pl/cms/fck/uploaded/strategia%20rozwoju%20miasta%20gdyni%202030_folder.pdf (accessed on 12 September 2020).

102. Strategia Rozwoju Krakowa. Tu chcę żyć. Kraków 2030. Available online: https://www.bip.krakow.pl/?dok_id=167&sub_dok_id=167&sub=uchwala&query=id%3D23155%26typ%3Du (accessed on 12 September 2020).
103. Strategia Rozwoju Miasta Torunia do roku 2020 z uwzględnieniem perspektywy rozwoju do 2028 roku. Available online: https://www.torun.pl/sites/default/files/pliki/1065_02.pdf (accessed on 12 September 2020).
104. #Warszawa 2030, Strategia. Available online: http://2030.um.warszawa.pl/wp-content/uploads/2018/06/Strategia-Warszawa2030-final.pdf (accessed on 12 September 2020).
105. Strategia Wrocław 2030. Available online: https://www.wroclaw.pl/rozmawia/strategia/Strategia_2030.pdf (accessed on 12 September 2020).
106. Zabrze Strategia Rozwoju Miasta. Available online: https://miastozabrze.pl/miasto/dokumenty-strategiczne/strategia-rozwoju-miasta-zabrze-2030/ (accessed on 12 September 2020).
107. Bydgoszcz 2030, Strategia Rozwoju. Available online: https://www.bydgoszcz.pl/fileadmin/%40strategia.bydgoszcz.pl/DOKUMENTY/Bydgoszcz_2030._Strategia_Rozwoju.pdf (accessed on 12 September 2020).
108. Plan Gospodarki Niskoemisyjnej dla m.st. Warszawy. Available online: http://infrastruktura.um.warszawa.pl/sites/infrastruktura.um.warszawa.pl/files/dokumenty/plan_gospodarki_niskoemisyjnej_dla_m.st_._warszawy.pdf (accessed on 18 September 2020).
109. Plan Gospodarki Niskoemisyjnej dla Gminy Miejskiej Kraków. 2017. Available online: https://www.bip.krakow.pl/_inc/rada/posiedzenia/show_pdfdoc.php?id=90606 (accessed on 18 September 2020).
110. Plan Gospodarki Niskoemisyjnej dla Miasta Łodzi. 2017. Available online: https://bip.uml.lodz.pl/files/bip/public/BIP_SS/WGK_zaktplan_20180122.pdf (accessed on 18 September 2020).
111. Plan Gospodarki Niskoemisyjnej dla Gminy Wrocław. 2017. Available online: https://www.wroclaw.pl/files/files/pdf-y/PGN_Wroclaw.pdf (accessed on 18 September 2020).
112. Plan Gospodarki Niskoemisyjnej dla Miasta Poznania. 2017. Available online: https://bip.poznan.pl/bip/uchwaly/uchwala-nr-lii-924-vii-2017-z-dnia-2017-07-11,69323/ (accessed on 18 September 2020).
113. Plan Gospodarki Niskoemisyjnej dla Miasta Gdańska. 2015. Available online: https://gdansk.xgcmy.pl/s1/d/20151164677/Plan-gospodarki-niskoemisyjnej-dla-Miasta-Gdanska.pdf (accessed on 18 September 2020).
114. Plan Gospodarki Niskoemisyjnej dla gminy miasto Szczecin. 2020. Available online: http://bip.um.szczecin.pl/files/42F6132C31634108BA29A4CECD84CF2E/576.pdf (accessed on 18 September 2020).
115. Plan działań na Rzecz Zrównoważonej Energii—Plan Gospodarki Niskoemisyjnej dla Miasta Bydgoszczy na lata 2014–2020+. 2016. Available online: https://www.czystabydgoszcz.pl/wp-content/uploads/2018/04/PGN_BYDGOSZCZ_Aktualizacja_2016.pdf (accessed on 18 September 2020).
116. Plan Gospodarki Niskoemisyjnej dla Miasta Lublin. 2020. Available online: https://lublin.eu/gfx/lublin/userfiles/_public/mieszkancy/srodowisko/energia/pgn/ii_aktualizacja_pgn_dla_miasta_lublin_2020.pdf (accessed on 18 September 2020).
117. Plan Gospodarki Niskoemisyjnej dla Miasta Białegostoku i Gmin Choroszcz, Czarna Białostocka, Dobrzyniewo Duże, Juchnowiec Kościelny, Łapy, Supraśl, Wasilków, Zabłudów do roku 2020. 2015. Available online: https://www.bialystok.pl/pl/dla_mieszkancow/ochrona_srodowiska/plan-gospodarki-niskoemisyjnej-dla-miasta-bialegostoku-i-gmin-choroszcz-czarna-bialostocka-dobrzyniewo-duze-juchnowiec-koscielny-lapy-suprasl-wasilkow-zabludow-do-roku-2020-1.html (accessed on 18 September 2020).
118. Plan Gospodarki Niskoemisyjnej dla Miasta Katowice. 2018. Available online: https://bip.katowice.eu/Lists/Dokumenty/Attachments/105488/sesja%20LII-1060-18.pdf (accessed on 18 September 2020).
119. Plan Gospodarki Niskoemisyjnej dla Gminy Miasta Gdyni na Lata 2015–2020. 2016. Available online: https://static.um.gdynia.pl/storage/__old/gdynia.pl//g2/2016_10/113049_fileot.pdf (accessed on 18 September 2020).
120. Plan Gospodarki Niskoemisyjnej dla Miasta Radomia. 2015. Available online: http://bip.radom.pl/ra/srodowisko/plany-i-programy/plan-gospodarki-niskoem/33950,Aktualizacja-planu-gospodarki-niskoemisyjnej-dla-miasta-Radomia.html (accessed on 18 September 2020).
121. Plan Gospodarki Niskoemisyjnej dla Gminy Miasta Toruń na Lata 2015–2020. 2016. Available online: http://bip.torun.pl/dokumenty.php?Kod=1246692 (accessed on 18 September 2020).
122. Kompleksowy Plan Gospodarki Niskoemisyjnej dla Miasta Sosnowiec. 2015. Available online: http://www.sosnowiec.pl/_upload/PGN%20Sosnowiec%2011.09.2015%20a.pdf (accessed on 18 September 2020).

123. Plan Gospodarki Niskoemisyjnej dla Miasta Rzeszowa. 2017. Available online: https://bip.erzeszow.pl/static/img/k02/SR/SR/Plan%20Gospodarki%20Niskoemisyjnej%20dla%20miasta%20Rzeszowa.pdf (accessed on 18 September 2020).
124. Plan Gospodarki Niskoemisyjnej dla Miasta Kielce. 2018. Available online: http://www.um.kielce.pl/gfx/kielce2/userfiles/files/gospodarka-niskoemisyjna/plan_gosp_niskoem_2018.pdf (accessed on 18 September 2020).
125. Plan Gospodarki Niskoemisyjnej dla Miasta Gliwice. 2019. Available online: https://bip.gliwice.eu/prawo_lokalne/uchwaly_rady_miasta,12840,1 (accessed on 18 September 2020).
126. Plan Gospodarki Niskoemisyjnej dla Miasta Zabrze. 2016. Available online: https://miastozabrze.pl/dla-mieszkancow/5457-2/plany-i-programy/plan-gospodarki-niskoemisyjnej-dla-miasta-zabrze/ (accessed on 18 September 2020).
127. Babbie, E. *Badania Społeczne w Praktyce*; Wydawnictwo Naukowe PWN: Warszawa, Poland, 2007.
128. Angrosino, M. *Badania Etnograficzne i Obserwacyjne*; Wydawnictwo Naukowe PWN: Warszawa, Poland, 2010.
129. Likert, R. A Technique for the Measurement of Attitudes. *Arch. Psychol.* **1932**, *140*, 5–55.
130. Krok, E. Budowa kwestionariusza ankietowego a wyniki badań. *Studia Inform. Pomerania* **2015**, *37*, 55–73.
131. Lisiecka, K.; Kostka-Bochenek, A. Case study research jako metoda badań naukowych. *Przegląd Organ.* **2009**, *10*, 5–23. [CrossRef]
132. Nowicki, M. *Nadchodzi era Słońca*; Wydawnictwo Naukowe PWN: Warszawa, Poland, 2012.
133. Urbaniec, K.; Mikulčić, H.; Rosen, M.A.; Duić, N. A holistic approach to sustainable development of energy, water and environment systems. *J. Clean. Prod.* **2017**, *155*, 1–11. [CrossRef]
134. Łucki, Z.; Misiak, W. *Energetyka a Społeczeństwo. Aspekty socjologiczne*; Wydawnictwo Naukowe PWN: Warszawa, Poland, 2010.
135. Szymańska, D.; Korolko, M.; Chodkowska-Miszczuk, J.; Lewandowska, A. *Biogospodarka w Miastach*; Wydawnictwo Naukowe Uniwersytetu Mikołaja Kopernika: Toruń, Poland, 2017.
136. DSRK, 2013, Polska Długookresowa Strategia Rozwoju Kraju. Trzecia fala Nowoczesności. Available online: http://kigeit.org.pl/FTP/PRCIP/Literatura/002_Strategia_DSRK_PL2030_RM.pdf (accessed on 17 September 2020).
137. Warszawa w Kierunku Smart City. Available online: https://content.knightfrank.com/research/1500/documents/pl/warszawa-w-kierunku-smart-city-april-2018-5463.pdf (accessed on 17 September 2020).
138. Strategia SMART_KOM, Czyli Mapa Drogowa dla Inteligentnych Rozwiązań w Krakowskim Obszarze Metropolitalnym. Available online: http://www.kpt.krakow.pl/wp-content/uploads/2015/03/raport_smart_kom_sklad_kor10_aktywny.pdf (accessed on 10 September 2020).
139. Smart City Wrocław. Available online: https://www.wroclaw.pl/smartcity/ (accessed on 10 September 2020).
140. Smart City Poznań. Available online: https://www.poznan.pl/mim/smartcity/ (accessed on 11 September 2020).
141. Rekomendacje do Strategii Rozwoju Miasta Kielce 2030+. W kierunku Smart City. Available online: http://www.um.kielce.pl/gfx/kielce2/userfiles/files/strategia/rekomendacje_kielce.pdf (accessed on 12 September 2020).
142. Chwieduk, D. Solar energy use for thermal application in Poland. *Pol. J. Environ. Stud.* **2010**, *19*, 473–477.
143. Zeibote, Z.; Volkova, T.; Todorov, K. The impact of globalization on regional development and competitiveness: Cases of selected regions. *Insights Reg. Dev.* **2019**, *1*, 33–47. [CrossRef]
144. Al-Badi, A.H.; Ahshan, R.; Hosseinzadeh, N.; Ghorbani, R.; Hossain, E. Survey of smart grid concepts and technological demonstrations worldwide emphasizing on the Oman perspective. *Appl. Syst. Innov.* **2020**, *3*, 5. [CrossRef]
145. Local Data Bank Statistical Poland. Available online: https://bdl.stat.gov.pl/BDL/dane/podgrup/tablica (accessed on 5 October 2020).
146. Członkowie Zwyczajni Stowarzyszenia Gmin Polska Sieć "Energie Cités". Available online: http://www.pnec.org.pl/pl/stowarzyszenie/czlonkowie-zwyczajni (accessed on 5 October 2020).
147. Które miasta dołączyły do grona zwycięzców konkursu Eco-Miasto? Available online: https://www.eco-miasto.pl/ktore-miasta-dolaczyly-do-grona-zwyciezcow-konkursu-eco-miasto (accessed on 5 October 2020).
148. Nugraha, A.R.; Subekti, P.; Romli, R.; Novianti, E. Public services optimizing through the communication and information technology application of local governments as an effort to form environmentally friendly smart city branding. *J. Phys. Conf. Ser.* **2019**, *1363*, 012055. [CrossRef]

149. Yigitcanlar, T.; Kamruzzaman, M.; Foth, M.; Sabatini-Marques, J.; da Costa, E.; Ioppolo, G. Can cities become smart without being sustainable? A systematic review of the literature. *Sustain. Cities Soc.* **2019**, *45*, 348–365. [CrossRef]
150. Kammen, D.M.; Sunter, D.A. City-integrated renewable energy for urban sustainability. *Science* **2016**, *352*, 922–928. [CrossRef]
151. Ortega, J.L.G.; Pérez, E.M. Spanish renewable energy: Successes and untapped potential. In *Renewable Energy Policy and Politics: A Handbook for Decision-Making*; Earthscan: London, UK, 2006; pp. 215–227.
152. Ajayi, O.O.; Ajanaku, K.O. Nigeria's energy challenge and power development: The way forward. *Energy Environ.* **2009**, *20*, 411–413. [CrossRef]
153. Castellanos, S.; Sunter, D.A.; Kammen, D.M. Rooftop solar photovoltaic potential in cities: How scalable are assessment approaches? *Env. Res. Lett.* **2017**, *12*, 125005. [CrossRef]
154. Kusch-Brandt, S. Underutilised Resources in Urban Environments. *Resources* **2020**, *9*, 38.
155. Sait, M.A.; Chigbu, U.E.; Hamiduddin, I.; de Vries, W.T. Renewable energy as an underutilised resource in cities: Germany's 'Energiewende'and lessons for post-brexit cities in the United Kingdom. *Resources* **2019**, *8*, 7. [CrossRef]
156. Szymańska, D.; Lewandowska, A. Biogas power plants in Poland—structure, capacity, and spatial distribution. *Sustainability* **2015**, *7*, 16801–16819. [CrossRef]
157. Jurasz, J.K.; Dąbek, P.B.; Campana, P.E. Can a city reach energy self-sufficiency by means of rooftop photovoltaics? Case study from Poland. *J. Clean. Prod.* **2020**, *245*, 118813. [CrossRef]
158. Moss, T.; Becker, S.; Naumann, M. Whose energy transition is it, anyway? Organisation and ownership of the Energiewende in villages, cities and regions. *Local Environ.* **2015**, *20*, 1547–1563.
159. Paul, F.C. Deep entanglements: History, space and (energy) struggle in the German Energiewende. *Geoforum* **2018**, *91*, 1–9. [CrossRef]
160. Abrao, R.R.; Paschoareli, D.; Silva, A.A.; Lourenco, M. Economic viability of installations of photovoltaic microgeneration in residencies of a smart city. In Proceedings of the 2017 IEEE 6th International Conference on Renewable Energy Research and Applications (ICRERA), San Diego, CA, USA, 5–8 November 2017; pp. 785–787.
161. IRENA. *Renewable Energy in Cities*; International Renewable Energy Agency (IRENA): Abu Dhabi, The United Arab Emirates, 2016. Available online: www.irena.org (accessed on 17 October 2020).
162. RENRenewables. *2019 Global Status Report*; REN21 Secretariat: Paris, France, 2019; Available online: http://www.ren21.net/gsr-2019/ (accessed on 17 October 2020).


**Publisher's Note:** MDPI stays neutral with regard to jurisdictional claims in published maps and institutional affiliations.

*Article*

# Smiling Earth—Raising Awareness among Citizens for Behaviour Change to Reduce Carbon Footprint

**Sobah Abbas Petersen [1,2,*], Idar Petersen [3] and Peter Ahcin [3]**

1 Department of Computer Science, Norwegian University of Science and Technology, 7491 Trondheim, Norway
2 Technology Management, SINTEF Digital, 7491 Trondheim, Norway
3 VINEGEN, 1500 Ljubljana, Slovenia; pinadoe@gmail.com (I.P.); Peter.Ahcin@gmail.com (P.A.)
* Correspondence: sap@ntnu.no

Received: 30 September 2020; Accepted: 11 November 2020; Published: 13 November 2020

**Abstract:** This paper describes Smiling Earth, a mobile app to increase citizens' awareness about their own carbon footprint, by integrating energy and transport-related data. The main aim of our work is to explore the ways in which Information and Communication Technologies could help raise awareness and educate and motivate citizens about their actions and their consequences on the environment. Smiling Earth provides feedback to users by visualising data about their daily activities with the aim to motivate citizens to change their behaviour to reduce their $CO_2$ emissions by adopting a healthier lifestyle. The value of the Smiling Earth for individuals, cities and communities is discussed. The feedback from an expert evaluation and how Smiling Earth could contribute to Positive Energy Districts are also discussed.

**Keywords:** behaviour change; carbon footprint: citizen motivation; sustainability

## 1. Introduction

The success of a smart and sustainable city development needs integrated solutions on energy, transport, service and governance, with the full involvement of multiple stakeholders, governments, enterprises and citizens. Recent research has focussed on a number of different technological solutions, such as sensors and Internet of Things [1], data analytics and Machine Learning, to monitor and control various aspects of a city. However, motivation and engagement of one of the most important stakeholders, the citizen, are not always in focus, and when they are, they often focus on one aspect only, such as energy [2] or transport [3].

A large part of the responsibility for meeting the United Nations (UN) Sustainable Development Goals (SDG), reducing energy consumption and climate change mitigation, has been taken on by municipalities. Policies and strategies are defined often at the national, European or global levels. In Europe, the Covenant of Mayors (CoM) initiative, which includes several cities, aims to identify how European policy could be enacted at the local authorities' level. They have made commitments to, for example, reduce their Carbon Dioxide ($CO_2$) emissions or transition to cleaner forms of energies, and set targets for them [4]. While this is a step to enacting policies, it is still a challenge to make individual citizens aware of their responsibilities for mitigating climate change.

A large part of greenhouse gas emissions which contribute to climate change come from transportation, electricity and heating. Households have multiple opportunities to reduce their carbon footprint, including energy conservation behaviours, the use of more energy-efficient vehicles or the installation of insulation products. Statistics from Norway [5] show that on average, 50% of households' incomes are spent on energy, transportation and housing. Transportation has been identified as one of the biggest contributors to $CO_2$ emissions [3]. Better energy and transportation management can lead to savings, and in Norway, transport is one of the sectors where most cuts in $CO_2$ emissions will be made in the near future [6].

The recent European focus on achieving Positive Energy Districts (PED) has increased the interest around energy efficiency, e.g., large European consortiums including private and public partners such as the EU H2020 +CityxChange (Positive Energy Exchange) project [7]. A PED is described in the European Strategic Energy Technology (SET-Plan) as "a district with annual net zero energy import, and net zero $CO_2$ emissions working towards an annual local surplus production of renewable energy" [8]. The PED concept is closely intertwined with the built environment and aims to achieve PED through energy efficiency in buildings, flexible energy consumption within districts and through local or regional supply of renewable energy. The SET-Plan identifies a broad spectrum of stakeholders in the context of PED and recognises citizens as a significant actor. In particular, the role of citizens in the energy exchange and energy trading activities, as prosumers and not only as consumers, is a major step towards citizen participation in achieving a PED. However, it is not clear how individual citizens could relate to contributing to a PED or a Zero Emission Neighbourhood (ZEN) [9] as a part of their lifestyle and daily activities.

Communicating the reality of climate change and its consequences has been a challenge and researchers and public authorities have tried to address this in many ways. Some of these include disseminating information, perhaps through mass media, assuming that information will inspire citizens to make good choices, attempts to motivate people by fear and visions of potential catastrophe and scientific framing of issues and impacts of climate change [10]. Increased energy literacy appears to be a popular strategy, by using persuasive technologies, data and visualisation to inform and make the users aware of their behaviour and the consequent emissions [11–13]. Informing consumers about energy consumption, either as direct or indirect feedback through displays and other means, has been identified as a means of raising awareness and influencing behaviour [14]. Games and game mechanics have also been popular strategies to motivate citizens as well as to provide feedback to them. Several types of incentives have also been identified, which address individuals as well as social aspects, and economic incentives appear to be popular [2]. Some of the shortcomings of existing communication and persuasion approaches about climate change are the lack of identification and support for the ways in which individuals and society in general could take action to address climate change and reflect on their actions and their consequences [10,15]. Furthermore, it is important for individuals to be able to see how their behaviours relate to the bigger context of climate change and sustainability.

The main aim of our work is to explore how Information and Communication Technologies (ICT) could help raise awareness, educate and motivate citizens about their actions and how they could impact the environment. We believe that an increased awareness among citizens about their own carbon footprint could contribute to help achieve the central aim of the Paris Agreement, which is to keep a global temperature rise below 2 °C [16]. Furthermore, raising awareness among citizens and providing feedback about how they could contribute to climate change through their daily activities are an important part of achieving PED and ZEN, to ensure alignment of citizens' behaviours and the current research activities within PED and ZEN, e.g., the Norwegian Research Centre on ZEN [9]. Thus, support towards changing citizens' behaviours towards a lower daily carbon footprint could make a significant impact on the environment. More specifically, in our work, we have focused on affecting citizens' behaviour to switch to less carbon-intensive modes of transport, such as electrical vehicles [17], walking or cycling, which would also contribute to their general well-being. We believe that this is a relevant step towards sustainable societies.

The Smiling Earth, a pervasive mobile app, is designed to visualise data about citizens' activities and to motivate citizens to change their behaviour to reduce their $CO_2$ emissions by adopting a healthier lifestyle. The app takes a broader perspective of $CO_2$ emissions, bringing together the transport and energy sectors. In addition, citizens (the users of the app) are provided information and feedback about how their choices of daily activities (e.g., transportation modes) could affect the environment as well as their own health. Emphasis was given to closing the gap between science or information and specific actions citizens could take. The initial version of Smiling Earth was designed to support individuals, which was further enhanced to support the social or community aspects. The Smiling Earth app was developed within the EU H2020 project, DESENT (grant agreement No. 693443).

The paper discusses the value of the Smiling Earth for individuals, cities and communities, through raising awareness about one's own carbon footprint and how citizens' actions through their daily choices could mitigate climate change. The paper also presents a qualitative evaluation of the relevance of bringing together transport- and energy-related data and visualising it and discusses the value of Smiling Earth to stakeholders outside of the DESENT project, based on semi-formal interviews.

The rest of this paper is organised as follows: Section 2 describes the DESENT project, Section 3 provides an overview of related work, Section 4 describes the research method and the design and development process for Smiling Earth, Section 5 describes the Transtheoretical Model of Behaviour Change, Section 6 describes the Smiling Earth mobile app, Section 7 describes the evaluation method, Section 8 presents the results of the evaluation and discusses them, Section 9 discusses and reflects on the feedback from the evaluation and Section 10 concludes the paper.

## 2. DESENT Project

Smart City-related research often focuses on silos or specific application domains. Smart City solutions and applications that leverage on the integration of several domains such as transport, energy, service and governance, and engage multiple stakeholders, such as governments, citizens, private and public entities, are important. The European H2020 project, DESENT (grant agreement No. 693443), focuses on providing a smart decision support tool for urban energy and transport, by developing innovative approaches and by utilising new and emerging technologies [18]. The DESENT consortium includes a diverse set of partners consisting of universities, research institutes, municipalities and private companies from three European countries (Austria, The Netherlands and Norway). DESENT was designed to support smart decision-making for policy makers and personalised services for citizens.

One of the main aims of DESENT is to integrate transportation and energy data to motivate citizens to change their daily habits by making better climate-related decisions. Energy companies and electricity grid owners are a part of the project consortium, who could leverage on this data and take the initiative to motivate behaviour change of their current and potential customers. The bigger context of using energy and transportation data to affect citizens' lives is shown in Figure 1.

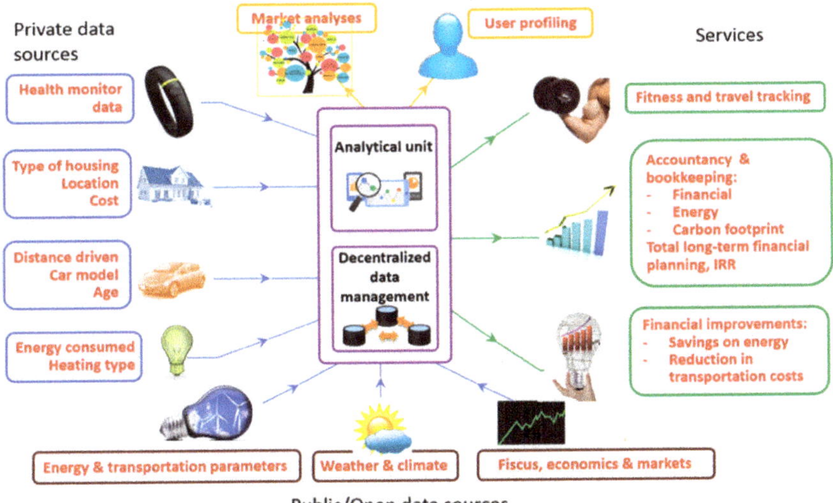


**Figure 1.** DESENT Smart Energy and lifestyle vision.

## 3. Related Work

Environmental sustainability and the use of ICT for informing, educating and affecting the behaviour of citizens have been the focus of several researchers. A variety of applications to promote the reduction of $CO_2$ through different activities have been developed, for example, EcoIsland is an application that promotes the reduction of $CO_2$ for individuals and families and it uses individual psychological incentives, social incentives and learning-related features [2,19]. Ducky is another application that quantifies, visualises and communicates everyday activities of individuals and communities that relate to climate, with the aim of increasing knowledge and awareness among citizens about their $CO_2$ emissions [13]. Both these examples focus on reducing energy consumption and use self-reporting as the means to capture users' behaviours. An example of an application that supports greener transportation is a gamified, in-car interface for motivating eco-driving, where feedback and rewards are used to motivate drivers to adopt eco-driving patterns for reducing emissions [20].

Mobile technologies and mobile phone apps have been popular in drawing the attention of individuals and engaging individuals on climate-related issues [21]. Such pervasive technologies address issues related to sustainability and climate change by encouraging greener transportation. A mobile application designed for tracking green transportation choices, UbiGreen, tracks an individual's transport options, such as walking or carpooling, and uses visual metaphors to provide feedback to the users [3]. In fact, metaphors are seen as a powerful means of communicating and providing feedback to the users, e.g., UbiGreen uses a metaphor from nature as the "wall paper". Based on the user's choices, the metaphor changes: one of the metaphors was a polar bear on an iceberg, where the iceberg increases in size if the individual makes greener transportation choices [22]. EcoIsland uses the metaphor of an island, whose water level rises if the users report activities that emit high levels of $CO_2$.

Several studies have shown that gamification can also be an effective means of engaging with diverse stakeholders and to improve knowledge among citizens, changes in public policy and behavioural change [23]. MUV (which stands for Mobility Urban Values) is a gamified mobile app that focuses on urban mobility, which is developed to encourage people to adopt sustainable mobility modes in the awareness of their potential role as agents of urban livability [24,25]. TrafficO2 is another mobile info-mobility decision support system, designed to encourage commuters towards more sustainable kinds of mobility by offering tangible incentives for more responsible choices [24]. These applications reward users based on their daily transportation choices, e.g., TrafficO2 rewards users with O2 points.

Most applications seem to focus on household-related energy consumption or transportation choices, while only a few attempt to provide a more integrated overview of emissions. Ducky takes a broad approach of supporting an eco-friendly lifestyle, e.g., by taking into account what one eats. An example that takes a broad perspective of the lifestyle is the Carbon Footprint Mobile App prototype described in [26], which is designed to help users to manage and maintain their daily carbon footprint by measuring their daily $CO_2$ emissions. The app acts as a self-tracking device and provides real-time data regarding an individual's daily carbon emissions. The app's design takes into account a user's $CO_2$ emissions related to clothing, food, household and transportation. The concept of betting is used to help users make a commitment to change their behaviours. While these applications provide data related to a user's carbon footprint and some provide feedback on the impacts to the environment, e.g., through the visual metaphors, we did not find applications that made citizens aware of the relationship of their actions towards achieving the global sustainability goals. Furthermore, several applications focused on making users aware of their actions through the self-reported or captured data and did not support users' estimate and plan to change their behaviour through informed sustainable choices.

## 4. Research Methods and Design Process

The overall research methodology that was used is the Design Science Research Method (DSRM) [17]. The context of our design or opportunity space has been the need to increase citizens' awareness about their own carbon footprints by integrating energy- and transport-related data.

The design and development process for Smiling Earth is illustrated in Figure 2. Leveraging on the access to an energy company, one of the partners in the consortium, we adopted a participatory design approach to design and develop the Smiling Earth pervasive mobile app. An agile iterative process and rapid prototyping, using the Proto.io tool, was followed to refine the concept and design of Smiling Earth, before it was developed. A series of formative evaluation workshops using the interactive prototype were conducted with energy experts, business and technical people from the energy company and a group of university students. The focus during the iterative design and formative evaluations was on usability and if the users understood the concepts. The design was improved based on the feedback from the studies. In the first phase of the development, Smiling Earth was focussed on affecting individuals' behaviour only. The feedback from summative evaluations, conducted with a group of university students and project partners, indicated that the concept of Smiling Earth was easy to understand, that it increased their awareness of their own carbon footprint and they are likely to use the app. Details of these evaluations are reported in References [27,28].

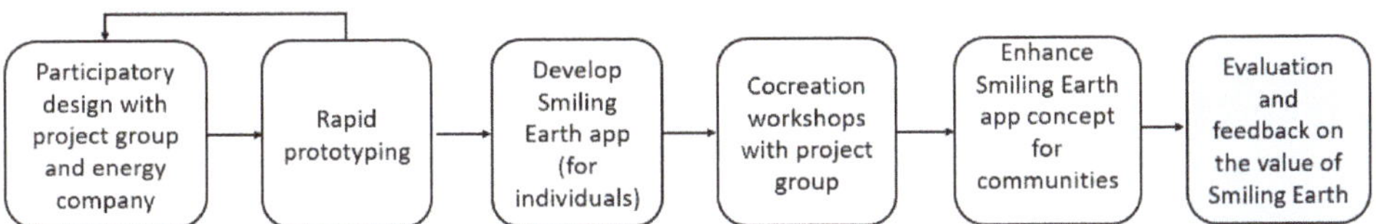


**Figure 2.** Overall design and evaluation process. The shaded box indicates the evaluation results reported in this paper.

Feedback from the project partners also indicated that it was desirable if Smiling Earth could affect the collective behaviour of a community of people, perhaps by including gamification or other motivational features. To enhance Smiling Earth from a mobile app for individuals to one for a community, we engaged with the project partners in a series of gamification and co-creation workshops, to design the enhanced functionalities in the app. The participants of the co-creation workshops included energy companies, transport researchers and people from two European municipalities from Austria and The Netherlands. The co-creation workshops used a set of social gamification cards, developed in the EU My Neighbourhood project, to co-design services in the public sector [29]. These workshops identified specific user groups, what their motivations may be and how they may be motivated to change their behaviour.

Sketches of the user interface for the enhanced Smiling Earth, to support a community of users, were then evaluated by people outside of the DESENT project, to evaluate if and how Smiling Earth could contribute to achieving PED and ZEN, and to obtain feedback on the benefits of Smiling Earth beyond the DESENT project. The section on evaluation and results will provide a detailed description of the evaluation method and process.

## 5. Behaviour Change

Designing technologies for affecting the behaviour change of individuals and communities require an understanding of what a change in behaviour means and how a change may be detected. Addressing an individual's attitude and their social norms are identified as important aspects of addressing behaviour change in the Theory of Planned Behaviour (TPB) by (Ajzen, 1991), as reported in References [30]. TPB and enhanced versions of this model have been used by researchers for affecting behaviour [31]. A comprehensive model for addressing the global environmental challenges and for modelling the behaviour of individuals was proposed by Kløckner, which is based on several existing theories [30]. Another popular model is the Transtheoretical Model (TTM), designed for individuals and intentional decision-making, and identifies six stages of changes an individual goes through [32]. It has been used in many health and behaviour change-related studies, e.g., to stop smoking [33]. It was developed through observations of behaviour in health/behaviour changes and recognises that individuals go through stages in changing their behaviour. Thus, this seemed an appropriate model to support changes in behaviour related to daily activities, which people have had for several years. Smiling Earth was designed as a smart decision-making tool for citizens and to affect behaviour change by taking action, and we have used TTM in our work.

The TTM model proposes six stages of behaviour change, where the first three stages propose that individuals contemplate on something before they take action, and eventually these actions become more regular and turn into a habit. The six stages of TTM are illustrated in Figure 3 and can be described as follows:

1. Precontemplation: individuals may not consider changing their behaviour, which may be due to a lack of awareness of the consequences of their behaviour.
2. Contemplation: individuals may intend to change their behaviour, but not immediately, and are more aware of the benefits of changing their behaviour.
3. Preparation: individuals are intending to take action in the near future.
4. Action: a change is observable, though individuals have to work hard to keep their new behaviour.
5. Maintenance: keeping the new behaviour is easier, but individuals may be tempted to revert back to their previous behaviour.
6. Termination: there is no more risk for individuals to revert back to their previous behaviour.

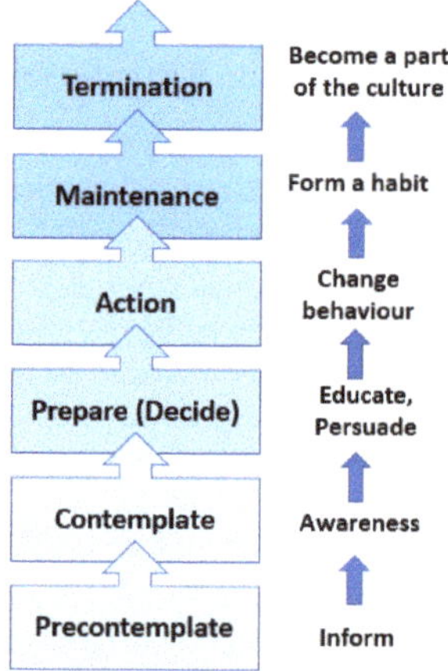


**Figure 3.** The Transtheoretical Model (TTM) and the behaviour change process.

We have adopted TTM as the conceptual framework for our design to support behaviour change through Smiling Earth. The first phase is to inform users and raise their awareness and then to persuade them to decide to change their behaviour. Following this, users must be supported to take action and continue to take action until it becomes a habit or a practice. Finally, support them to take the desired actions until it becomes a part of their lives.

One of the advantages of using stage models such as TTM is that it is easier to design technologies with the intent to address one or more of the stages. By understanding behaviour changes as stages of changes, it helps to recognise if the technology (or the intervention) has the potential to affect the behaviour in any way, e.g., even if the desired end result has not been achieved, rather than concluding that it has no effect, an evaluation using the TTM could discover that the intervention had encouraged stage progression, if not actual behavioural change [34].

One of the weaknesses of this model with respect to the evolution of Smiling Earth to support communities of people is that TTM was designed to support individuals and does not take into account the social context and thus, for our further work, we will consider other models that are better suited for collective behaviour change.

## 6. Description of Smiling Earth

The main concepts that have influenced the design of Smiling Earth were $CO_2$ emissions, environmental impact, health benefits and economic profit. The Graphical User Interface (GUI) for the main screen is designed to reflect these concepts and draw the user's attention to $CO_2$ emissions and its impact on the world we live in (see Figure 4.). We have used the earth as a metaphor, which appears in the centre of the screen. This metaphor is used to create an emotional attachment [22] and to show the impact of a high carbon footprint on the earth (global warning) which may be caused by the user's actions.

**Figure 4.** Smiling Earth concept, Graphical User Interface (GUI) and the Earth metaphor. (**a**) shows a very happy earth due to low $CO_2$ emissions; (**b**) shows a less happy face as the $CO_2$ emissions increase; (**c**) shows a happy earth due to low $CO_2$ emissions from an electrical vehicle and (**d**) shows a very miserable earth due to a high level of $CO_2$ emissions.

Smiling Earth tracks a user's movements using the Global Positioning System (GPS) and Google's Activity Recognition Application Programming Interface API to determine if the user is walking, cycling or in a vehicle. Furthermore, the household energy consumption is determined by energy classification of buildings, which are available through the public records in Norway and Austria. Users can provide information related to their profile, such as the energy source(s) in their houses (which is at the household level), and the type(s) of vehicle(s) they own. These data are used to estimate a user's daily energy consumption, and thus an estimation of their daily carbon footprint. The estimations are based on statistical data [35] and analyses conducted by SINTEF Energy and the DESENT partners. Since the main aim of Smiling Earth was to raise awareness and to support an individual to relate the impacts of their own activities to the 2 °C target, the focus was not on collecting real-time data.

The main screen for Smiling Earth shows the dashboard for the app, which is shown in Figure 4a–d. These screens show the estimates of daily values of $CO_2$ emissions from domestic heating and transportation. The circular indicator provides information about the current value of $CO_2$ emissions in kg of $CO_2$. The daily value is indicated by the number in the circle. The circular progress bar for each circular indicator shows the estimated value relative to the maximum allowed level of emissions, aimed to motivate the users to keep their $CO_2$ emissions below this maximum amount. The daily goal for the $CO_2$ limit is 4 kg $CO_2$ for the carbon footprint. The screenshots in Figure 4 show three possible states (out of five) for the earth metaphor: Figure 4a shows a very happy earth due to low emissions with respect to the maximum level, Figure 4b shows a happy face, although less happy than Figure 4a as the $CO_2$ emissions are closer to the maximum than in Figure 4a,d shows a very miserable picture of the earth because of the high level of emissions, which is much greater than the maximum level.

The colours blue and green have been used in the Graphical User Interface (GUI): blue indicates estimated values that are related to housing (such as the household heating and electricity consumption) and green indicates estimated values which are related to transportation, such as walking, cycling or electrical vehicles (EVs).

Figure 4c,d show estimated values of $CO_2$ emissions, based on the user's selection of a transportation or heating mode. The symbols on the bottom of the screen are "estimation buttons", from left to right, for solar panels, walking, cycling and an electrical vehicle. By selecting an "estimation button", the relevant values will be shown. For example, in Figure 4c, the estimated value for $CO_2$ emission by selecting an electrical vehicle is shown as 0.0 kg. Similarly, the estimated value for $CO_2$ emission from walking while either sometimes driving a non-electrical vehicle or using conventional heating could lead to a higher $CO_2$ emission, as shown in Figure 4d. The estimation capability is also used to support users to set short- or long-term goals, so that they could estimate how much they could reduce their carbon footprint, save (money) or burn calories over time. This is aimed to support users' plan for a bigger transition over time, through their daily activities.

A menu on the top left of the screen shows the users their estimated $CO_2$ emissions and other estimated values such as the amount of money they could save (in Norwegian krones) and calories burnt, over a period of time, e.g., a week, month or year. The visualisation of $CO_2$ emissions for a week and for a month are shown in Figure 5. The red horizontal line marks the maximum limit of $CO_2$ emissions, to encourage users to keep their total below that level. The blue and green colours indicate the emissions due to household actions or transportation actions, e.g., Figure 5a shows that 5 kg of $CO_2$ out of the total of 8 kg were from household activities. Bar graphs are used to visualise the data for a week while the continuous graph shown in Figure 5b seemed a better visualisation of the data for a longer time period, such as a month or a year.

**Figure 5.** Smiling Earth—$CO_2$ emissions data visualisation. (**a**) shows $CO_2$ for a week and (**b**) shows $CO_2$ emissions for a month.

A detailed description of Smiling Earth is available from Reference [36]. The Smiling Earth is designed to run in the background on a mobile phone, and the sensors on the phone will detect a user's movements to estimate the energy consumption from mobility. The app is designed to minimise the information that a user has to provide and utilises data that are openly available, e.g., in Norway, it is possible to access information about vehicles registered under the name of each person and in

Austria, the age of a building provides the relevant information for estimating the energy consumption of the building.

*Smiling Earth for Communities*

Supporting a community of users was considered as an important element to motivate users to continue using the app and to achieve behaviour change, i.e., continue taking action to form new habits and making those a part of their daily lives, as illustrated in Figure 3. Thus, based on the input from the co-creation workshops with the project partners, the design of Smiling Earth was enhanced to support the social and community aspects. Some of the ideas that emerged from the co-creation workshops are described below:

- Motivate young people to start using Smiling Earth and continue using it, by making them feel a sense of belonging to the community. Users could be encouraged by recognising their use of the app by showing progression levels and by recognising frequent users as Heroes (within the (user) community)).
- Motivate older users to continue using Smiling Earth. The means of motivating older and younger users may be different. The sense of achievement and belonging may motivate older users. Hence, older users could be given a "gift" through the app, for walking rather than using the car, thus also promoting a healthier lifestyle.
- Motivate car owners to use their cars less by providing a visualisation of how they fare among other users in the community and visualising this. Users could also be motivated by rewarding them, for example with "Earth coins" or a digital or virtual currency that could be exchanged to perhaps a monetary value, a free bus ticket or a voucher, to get them onboard.

The design of the new features aimed to motivate all users while avoiding disappointing any users. The social game mechanic, Leaderboard, was incorporated to show users that were leading in many ways, e.g., one possibility could be based on the carbon footprint of the users while another possibility could be the number of Earth coins (analogous to points) earned by a user; see Figure 6. Users are able to give away some of their Earth coins to other users as a means of encouraging peers.

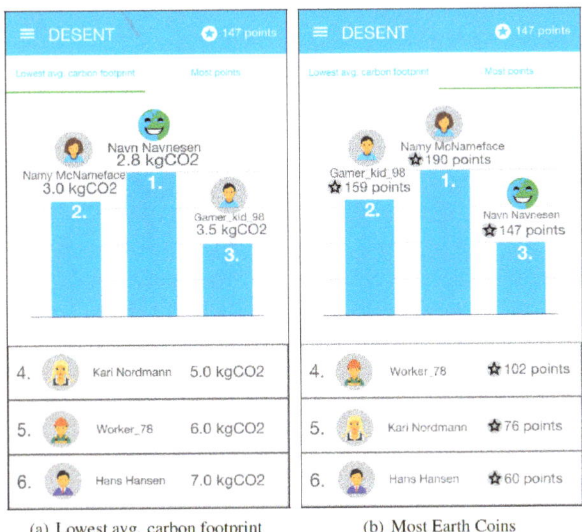


(a) Lowest avg. carbon footprint (b) Most Earth Coins

**Figure 6.** Smiling Earth for Communities showing examples of different ways of showing Leaderboards; (**a**) shows a leaderboard for lowest $CO_2$ emissions and (**b**) shows a leaderboard for earth coins collected.

Concepts such as Earth coins were introduced to encourage long-term goals and planning to achieve a goal over time that could lead to a bigger impact on the daily carbon footprint, such as investment in solar panels or an electrical vehicle. The concept of challenges was also supported so that users could set goals or challenges for themselves, or by a community leader, so that users could track their progress in the behaviour change. A detailed design of the gamified Smiling Earth for communities of users is available from Reference [37].

## 7. Evaluation

The agile design and development approach included several formative evaluations, consisting of the evaluation of the concept, user interface and usability of the app. A summative evaluation of Smiling Earth for individuals was conducted with several university students, using a questionnaire and follow-up interviews. The main feedback from those evaluations indicated that the concepts were interesting and motivating [27,28].

The results described in this paper focus on the value of Smiling Earth in an urban community context, for raising awareness among individuals and communities about their own carbon footprint. The Expert Evaluation Method [38] was adopted as comments, views and feedback from experts were important. Thus, the evaluation was conducted as a semi-structured interview. The interviews were conducted online, due to COVID-19 restrictions and the geographical distribution. The evaluation was done in two parts: the first part evaluated the value of Smiling Earth for affecting individuals and the second part evaluated a gamified version of Smiling Earth to affect the behaviour of a community or a group of people. A short presentation describing Smiling Earth, the main concepts and the user interface were shared among the participants prior to the interview. Each interview lasted between 1 and 2 h, depending on the discussions related to the open-ended questions.

As the institutions where the authors work are currently involved in two large research projects related to PED and ZEN, namely EU H2020 Positive City Exchange project [7] and Zero Emission Neighbourhoods [9], the experts for the evaluation were selected from the participants of these projects. Four experts participated in the evaluation: one is a researcher, responsible for working with the pilot neighbourhoods in the FME-ZEN project in Trondheim, the second works with citizen engagement in Trondheim Municipality, as a part of the +CityxChange project, the third is a researcher from University of Limerick working with citizen engagement, as a part of the +CityxChange project and the fourth is responsible for research projects in the Limerick City Council, Ireland. The two experts from Limerick were interviewed together as that was the best option to meet their availability.

The evaluations consisted of a short questionnaire as well as open ended questions, opening up for the free expression of the participants. The questionnaire had a set of statements and participants were asked to indicate to what degree they agreed with the statements. A Likert scale from 0 to 4 was used, where 4 indicated strong agreement and 0 indicated strong disagreement.

## 8. Results

This section presents the results of the evaluation. The first subsection addresses the evaluation of the Smiling Earth app for raising awareness among individual citizens and the second subsection presents the evaluation of a gamified Smiling Earth app for raising awareness of a community of citizens. The participants of the evaluations are referred to as P1, P2, P3 and P4 in the graphs shown in Figures 7 and 8.

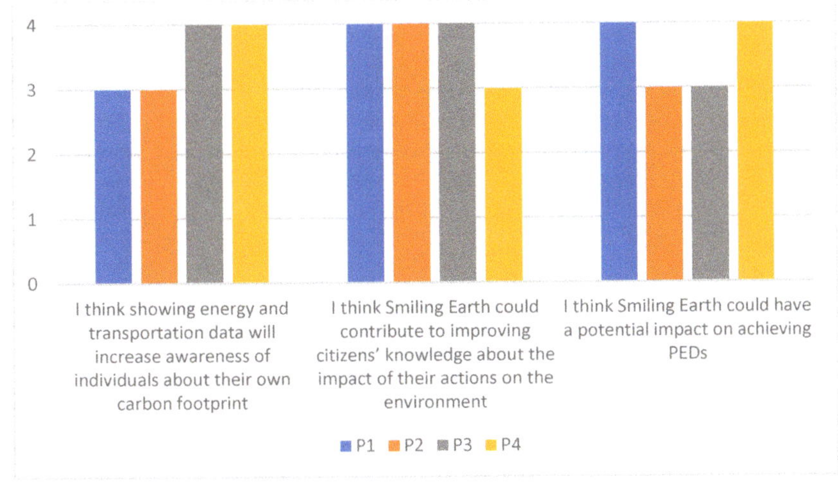


**Figure 7.** Evaluation results—Smiling Earth for raising awareness among individual citizens.

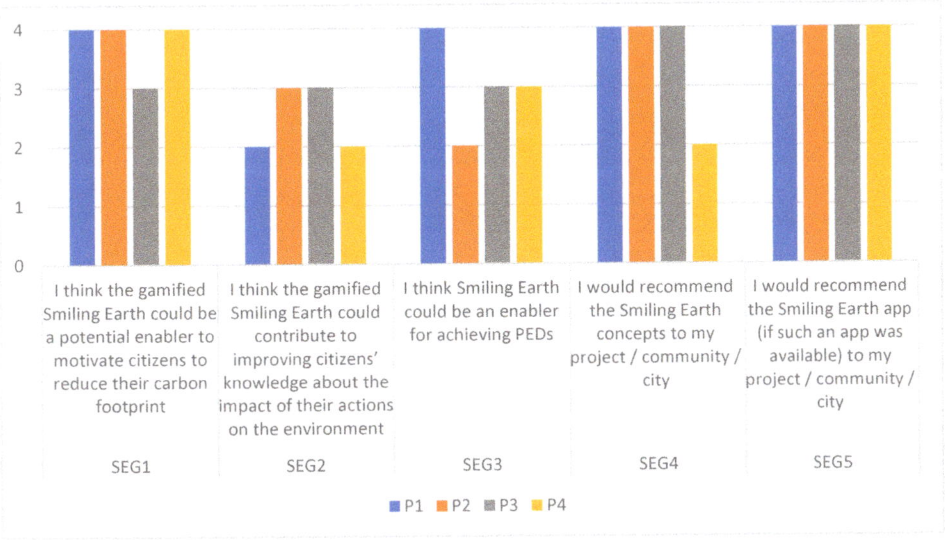


**Figure 8.** Gamified Smiling Earth for raising awareness in communities.

### *8.1. Smiling Earth for Raising Awareness among Individual Citizens*

The results for the short questionnaire are shown in Figure 7 where the three statements are shown on the horizontal axis and the Likert scale and the participants' responses are shown on the vertical scale. The average scores for the statements were 3.5 and 3.75, indicating strong agreement to the statements by all 4 participants. All participants agreed to the relevance of integrating energy and transport data and the benefits of visualising this. The relationship between an individual's carbon footprint and the current focus on PED is often not clear and this will be discussed in more detail in Section 9.

The semi-structured interview was based on three additional questions and the participants' responses to these are summarised for each question.

"What do you expect as the potential benefit for your project/community/city?" It appears that a lot of the current focus in projects places the responsibility of climate change on the buildings and the energy sector and the municipalities, and not on individual citizens. This has been a challenge in projects to engage citizens in the pilot neighbourhoods and cities. The Smiling Earth app is seen as a means to help raise awareness among citizens as well as among other stakeholders, e.g., the education sector and the school system. The visualisation helps people relate to how they move around and the subsequent energy consumption. Relating these to daily activities could also serve as daily reminders, e.g., for parents, who perhaps have to drive their children to various activities, to save time. It raises awareness of an individual's impact on the climate. It was noted that most citizens do not have a feeling about what energy is or what a unit of energy really is and Smiling Earth could help citizens relate to and understand what a unit of energy is and how that relates to individual energy consumption. It was also noted that it was good to have calories included as a part of the estimations related to daily choices.

"Who would have the most value from the app—individual citizen, Community, city? And what is the value?" The feedback on a citizen's choice and personal behaviour provided by Smiling Earth can be of value for citizens. In particular, most plans by municipalities or at a higher level such as the Paris Agreement, do not integrate citizens. This is where Smiling Earth could help to breakdown climate change activities to the lowest level for citizens to relate to them. Smiling Earth could be of value for cities and neighbourhoods by identifying patterns of transportation by citizens. The data from the app could provide relevant data for city planning and transportation planning. Two participants noted that the benefit for the city or a community depends on the number of people using it. One participant noted that the benefit is really for the earth and everyone as anything that helps mitigate climate change is good for all.

Participants were asked for other general comments and suggestions. One participant noted that this app was perhaps not for the older generations, but for the more data-literate users. It was also noted that the next step would be to leverage on the data that is gathered through the app and to enhance it towards supporting a community of users around it.

### 8.2. Gamified Smiling Earth for Raising Awareness in Communities

This part of the evaluation consisted of five statements where the participants had to indicate their agreement using the Likert scale. The results are shown in Figure 8 where the five statements are shown on the horizontal axis and the Likert scale and the participants' responses are shown on the vertical scale. The average scores for three of the five questions were between 3.5 and 4, indicating a strong agreement with the statements. The statement "I think the gamified Smiling Earth could contribute to improving citizens' knowledge about the impact of their actions on the environment" has an average score of 2.5, indicating neither agreement nor disagreement. The feedback from two of the participants was that "gamifying itself would not make a difference", and that it was not necessarily the citizens' knowledge that could be improved, but their motivation. The statement "I think Smiling Earth could be an enabler for achieving PEDs" has an average score of 3, indicating agreement with the statement. However, this is the statement that has most variance among the responses. The discussions that followed noted that Smiling Earth is focused on individuals and the value of the app to PED could depend on how one defines a PED as well as other factors, which will be discussed in the Section 9. For the statement "I would recommend the Smiling Earth concepts to my project/community/city", one of the participants, who worked in a city council, shared that he would not bring a concept to the citizens unless it was proven to work, as a premature introduction to citizens could have negative consequences.

## 9. Discussion

Smiling Earth integrates energy- and transport-related data to visualise an individual's energy consumption during their daily activities. The visualisations and graphics were described as "easy and intuitive" and the earth metaphor was seen as potentially having a high impact among the users. It is seen as a means of reaching more people than through most of the current approaches. The feedback from the evaluations confirm that the Smiling Earth could play an important role in raising awareness among individuals to help them understand what energy is, what a unit of energy means and most importantly, to relate their energy consumption to the bigger picture of global climate change. One of the challenges in raising awareness among citizens is relating high-level goals such as the UN SGDs or the aims of the Paris Agreement to individuals' actions, and to create a sense of responsibility at the citizen level. Furthermore, it is often difficult to understand the science-based facts and information, e.g., what a 2 °C temperature rise means. Increasing the knowledge and awareness among citizens is seen as a necessary step in achieving the global goals. Smiling Earth could be valuable in supporting this step for getting a broader range of stakeholders and citizens onboard and contributing to achieving the global sustainability goals. More importantly, it can play an important role in closing the science–action gap [10], to help citizens take appropriate action towards mitigating climate change.

While Smiling Earth was initially designed to support individuals, the value of it in engaging citizens and communities of citizens was identified by all the participants. Communities or a city as a collective of people could benefit enormously from using Smiling Earth if a great number of people use the app. One participant emphasised the value of users "networked as a community" through the app, engaging several users to reduce their carbon footprints and change their behaviours. The gamified Smiling Earth to support communities or groups of users was a step in this direction. The importance of the design of the gamification was raised in the discussions as gamification could sometimes lead to a counter effect or undesired behaviour. For example, game mechanics such as a competition among users could lead to undesired behaviour such as cheating and promote competition rather than collaboration. The challenges of designing good social game mechanics have been discussed in other projects, too [39]. Therefore, the choice of game mechanics to engage and motivate collective effort and collective behaviour change of a community of users, where individual users motivate, encourage and support other users to change their behaviour, are desired.

The possibility for users to be able to set goals or that challenges could be set for the users were seen as benefits, and it is important to set goals that are achievable for the users, so that they are motivated rather than disappointed. The importance of creating a feeling of self-efficacy, the ability to perform the necessary act [30], was highlighted.

Smiling Earth was not designed to take into account geographical boundaries, but to focus on $CO_2$ emissions and citizens' carbon footprints. Thus, discussions about how Smiling Earth could contribute to a PED or a ZEN highlighted interesting issues. The definition of a PED focusses on a geographical area or a district and energy use and consumption [8], and focusses less on $CO_2$ emission. However, the participants of the evaluation believed that Smiling Earth will have an impact on PED. In fact, it could provide the "missing link" between a PED and an individual's or citizens' actions. One interesting issue was that some people lived and worked in two different geographical locations (districts, neighbourhoods or cities), and the different areas offer different possibilities for transportation or energy. Thus, people's choices are often influenced or constrained by the choices available for them. For example, two of the participants lived in less urban areas and worked in the city, necessitating them to drive to work. So, although they may have a low carbon footprint from all their other activities (e.g., live in a passive house), their net carbon footprint may still be high, w.r.t to the daily goal for the $CO_2$ limit, which is 4 kg $CO_2$. The evaluation of Smiling Earth stimulated a number of interesting questions, such as could people have different climate change-related behaviours in different districts and what would this imply for districts and cities? An important revelation is that Smiling Earth can stimulate important discussions around people's energy-related behaviours, thereby indirectly raising people's awareness about climate change and their own actions. Furthermore, these discussions provided insights into the potential of Smiling Earth and how it could be adapted for specific needs of citizens, cities or geographical areas.

Another interesting point was how one would define a city in this context: "is a city an institution or a collection of people"? This is an important question when cities (or municipalities) consider the adoption of technologies such as Smiling Earth. Indeed, this is an important design consideration for designers of technologies and services to support citizen engagement towards climate change and sustainability.

## 10. Conclusions

This paper described Smiling Earth, a mobile app to increase citizens' awareness about their own carbon footprint, by integrating energy- and transport-related data. The main aim of our work was to explore the ways in which ICT could help raise awareness and educate and motivate citizens about their actions and their consequences on the environment. Smiling Earth provides feedback to users by visualising data about their activities with the aim to motivate citizens to change their behaviour to reduce their $CO_2$ emissions by adopting a healthier lifestyle.

The focus of this paper has been to validate the concept of Smiling Earth and to evaluate its value beyond the project it was developed in. Hence, the paper reported an expert evaluation, based on semi-structured interviews, to establish the potential value of Smiling Earth to individuals, cities and neighbourhoods. The feedback from the evaluations were positive and identified several benefits to citizens and communities of people. One of the most important values that was identified was that Smiling Earth could be the missing link between global or national sustainability goals and policies and the citizens. The evaluation also identified interesting discussion points such as how individuals' behaviours relate to PED or specific geographical areas. Examining the role of technologies such as Smiling Earth from the different perspectives of an individual, a community or a city stimulated interesting discussions about what a city could be: is it an institution or a collective of people?

The evaluation increased our awareness about the different perspectives of the stakeholder groups and their expectations, providing a better insight and stimulating design ideas. This also helped us identify some of the weaknesses in the design of the statements used in the qualitative evaluations. Our future work will include the enhancement of Smiling Earth to motivate communities of users towards behaviour change. This will include further studies with a broader set of user groups, such as people of different age groups, people living in urban and rural areas, different transportation needs and mobility patterns.

**Author Contributions:** Conceptualisation of Smiling Earth, P.A. and I.P.; methodology, S.A.P., P.A. and I.P.; validation, S.A.P.; formal analysis, S.A.P. and I.P.; writing, S.A.P.; writing—reviewing and editing, I.P. All authors have read and agreed to the published version of the manuscript.

**Funding:** This research was funded mostly through the EU H2020 DESENT project, under grant agreement No. 693443, where SINTEF Energy, Norway, was the project partner. The first author is partly supported by FME-ZEN project, financed by the Norwegian Research Council, and the +CityxChange project, Grant agreement 824260.

**Acknowledgments:** This work has been conducted as a part of the EU H2020 DESENT project, under grant agreement No. 693443, where SINTEF Energy, Norway, was the project partner, represented by the 2nd and 3rd authors. The authors would like to thank the Masters' students from NTNU, Norway, Celine Minh and Ragnhild Larsen for implementing Smiling Earth. The authors are also grateful for the participants of the evaluations reported in this paper: Daniela Baer, SINTEF Community (from the FME-ZEN project), Cole Matthew Grabinsky, Trondheim Municipality, Norway, Helena Fitzgerald, University of Limerick and Terence Connolly, Limerick City Council, Ireland. The latter three are also participants of the EU H2020 SCC +CityxChange project.

**Conflicts of Interest:** The authors declare no conflict of interest.

## References

1. Rahwan, T.; Michalak, T.; Elkind, E.; Faliszewiski, P. Constrained Coalition Formation. In Proceedings of the Twenty-Fifth AAAI Conference on Artificial Intelligence, San Francisco, CA, USA, 7–11 August 2011.
2. Shiraishi, M.; Washio, Y.; Takayama, C.; Lehdonvirta, V.; Kimura, H.; Nakajima, T. Using individual, social and economic persuasion techniques to reduce CO2emissions in a family setting. In Proceedings of the 4th International Conference on Theory and Practice of Electronic Governance—ICEGOV '10, Bogota, Colombia, 10–13 November 2009.

3. Froehlich, J.; Dillahunt, T.; Klasnja, P.V.; Mankoff, J. UbiGreen: Investigating a mobile tool for tracking and supporting green transportation habits in CHI '09. In Proceedings of the 27th International Conference on Human Factors in Computing Systems, Boston, MA, USA, 4–9 April 2009.
4. Kona, A.; Bertoldi, P.; Kılkış, Ş. Covenant of Mayors: Local Energy Generation, Methodology, Policies and Good Practice Examples. *Energies* **2019**, *12*, 985. [CrossRef]
5. Statistics Norway. Survey of Consumer Expenditure. 2012. Available online: https://www.ssb.no/en/inntekt-og-forbruk/statistikker/fbu (accessed on 26 June 2017).
6. Granell, C.; Havlik, D.; Schade, S.; Sabeur, Z.; Delaney, C.; Pielorz, J.; Usländer, T.; Mazzetti, P.; Schleidt, K.; Kobernus, M.; et al. Future Internet technologies for environmental applications. *Environ. Model. Softw.* **2016**, *78*, 1–15. [CrossRef]
7. +CityXChange. Positive City Exchange. Available online: http://cityxchange.eu/ (accessed on 19 May 2019).
8. SET-Plan ACTION n.3.2 Implementation Plan. Available online: https://setis.ec.europa.eu/system/files/setplan_smartcities_implementationplan.pdf (accessed on 16 September 2016).
9. FME-ZEN. Zero Emission Neighbourhoods Research Centre. Available online: https://fmezen.no/ (accessed on 29 September 2020).
10. Moser, S.C.; Dilling, L. *Communicating Climate Change: Closing the Science-Action Gap*; Oxford University Press: London, UK, 2011; pp. 161–176.
11. BeAware: Boosting Energy Awareness. Available online: http://www.energyawareness.eu/beaware/ (accessed on 26 June 2017).
12. Brewer, R.S.; Lee, G.E.; Johnson, P.M. The Kukui Cup: A Dorm Energy Competition Focused on Sustainable Behavior Change and Energy Literacy. In Proceedings of the 44th Hawaii International Conference on System Sciences, Institute of Electrical and Electronics Engineers (IEEE), Kauai, HI, USA, 4–7 January 2011; pp. 1–10.
13. Glogovac, B.; Simonsen, M.; Solberg, S.S.; Löfström, E. Ducky: An Online Engagement Platform for Climate Communication. In Proceedings of the 9th Nordic Conference CHI, Gothenburg, Sweden, 23–27 October 2016.
14. Zangheri, P.; Serrenho, T.R.; Bertoldi, P. Energy Savings from Feedback Systems: A Meta-Studies' Review. *Energies* **2019**, *12*, 3788. [CrossRef]
15. Brynjarsdottir, H.; Håkansson, M.; Pierce, J.; Baumer, E.; Disalvo, C.; Sengers, P. Sustainably Unpersuaded. In Proceedings of the SIGCHI Conference on Human Factors in Computing Systems, Boston, MA, USA, 5–10 May 2012; pp. 947–956.
16. United Nations Framework Convention on Climate Change UNFCCC. What Is the Paris Agreement? Available online: https://unfccc.int/process-and-meetings/the-paris-agreement/what-is-the-paris-agreement#:~{}:text=The%20Paris%20Agreement\T1\textquoterights%20central%20aim,further%20to%201.5%20degrees%20Celsius (accessed on 29 September 2020).
17. Hevner, A.R.; March, S.T.; Park, J.; Ram, S. Design Science in Information Systems Research. *MIS Q.* **2004**, *28*, 75. [CrossRef]
18. DESENT. Smart Decision Support System for Urban Energy and Transportation. Available online: www.jpi-urbaneurope.eu/desent (accessed on 14 November 2019).
19. Takayama, C.; Lehdonvirta, V.; Shiraishi, M.; Washio, Y.; Kimura, H.; Nakajima, T. ECOISLAND: A System for Persuading Users to Reduce CO2 Emissions. In Proceedings of the 2009 Software Technologies for Future Dependable Distributed Systems, Institute of Electrical and Electronics Engineers (IEEE), Tokyo, Japan, 17 March 2009; pp. 59–63.
20. Inbar, O.; Seder, T. Driving the Scoreboard: Motivating Eco-Driving Through In-Car Gaming. In Proceedings of the CHI 2011, Vancover, BC, Canada, 7–12 May 2011.
21. Sagawe, A.; Funk, B.; Niemeyer, P. Modeling the Intention to Use Carbon Footprint Apps. In *Quality, IT and Business Operations*; Springer Science and Business Media LLC: Berlin, Germany, 2016; pp. 139–150.
22. Dillahunt, T.; Becker, G.; Mankoff, J.; Kraut, R.E. Motivating Environmentally Sustainable Behavior Changes with a Virtual Polar Bear. In Proceedings of the 6th International Conference on Pervasive Computing, Sydney, Australia, 19 May 2008.
23. Rajanen, D.; Rajanen, M. Climate Change Gamification: A Literature Review. In Proceedings of the GamiFIN 2019 Conference, Levi, Finland, 8–10 April 2019.
24. Caroleo, B.R.D.; Morelli, N.; Lissandrello, E.; Vesco, A.; Di Dio, S.; Mauro, S.; Dio, D. Measuring the Change Towards More Sustainable Mobility: MUV Impact Evaluation Approach. *Systems* **2019**, *7*, 30. [CrossRef]

25. Di Dio, S.; Lissandrello, E.; Schillaci, D.; Caroelo, B.; Vesco, A.; D'Hespel, I. MUV: A Game to Encourage Sustainable Mobility Habits in Games and Learning Alliance. In Proceedings of the 7th International Conference GALA 2018, Palermo, Italy, 5–7 December 2018.
26. Fazeli, S.M. Smart City: A Prototype for Carbon Footprint Mobile App. In *Electrical Engineering*; KTH: Stockholm, Sweden, 2014.
27. Petersen, S.A.; Petersen, I.; Ahcin, P. Smiling Earth—Raising Citizens' Awareness on Environmental Sustainability. In *Business Intelligence*; Springer Science and Business Media LLC: Berlin, Germany, 2020; pp. 45–57.
28. Petersen, S.A.; Ahcin, P.; Petersen, I. Smiling Earth—Citizens' Awareness on Environmental Sustainability Using Energy and Transport Data. In *Public-Key Cryptography—PKC 2018*; Springer Science and Business Media LLC: Berlin, Germany, 2019; pp. 459–465.
29. Oliveira, M.; Petersen, S.A. Co-design of Neighbourhood Services Using Gamification Cards. In Proceedings of the Public-Key Cryptography–PKC 2018, Rio De Janeiro, Brazil, 25–29 March 2018.
30. Klöckner, C.A. A comprehensive model of the psychology of environmental behaviour—A meta-analysis. *Glob. Environ. Chang.* **2013**, *23*, 1028–1038. [CrossRef]
31. Barbarossa, C.; Beckmann, S.C.; De Pelsmacker, P.; Moons, I.; Gwozdz, W. A self-identity based model of electric car adoption intention: A cross-cultural comparative study. *J. Environ. Psychol.* **2015**, *42*, 149–160. [CrossRef]
32. Prochaska, J.O.; Velicer, W.F. The Transtheoretical Model of Health Behavior Change. *Am. J. Health Promot.* **1997**, *12*, 38–48. [CrossRef] [PubMed]
33. LaMorte, W.W. Behaviour Change Models—The Transtheoretical Model (Stages of Change). Available online: https://sphweb.bumc.bu.edu/otlt/mph-modules/sb/behavioralchangetheories/BehavioralChangeTheories6.html (accessed on 30 October 2020).
34. Howell, R.A. Using the transtheoretical model of behavioural change to understand the processes through which climate change films might encourage mitigation action. *Int. J. Sustain. Dev.* **2014**, *17*, 137. [CrossRef]
35. Statistics Norway. Emissions of Greenhouse Gases, 1990–2015, Final Figures. Available online: https://www.ssb.no/en/natur-og-miljo/statistikker/klimagassn/aar-endelige/2016-12-13?fane=tabell&sort=nummer&tabell=287212 (accessed on 28 June 2017).
36. Minh, C. *Mobile App: Integrated Financial, Energy and Transport Services*; Norwegian University of Science and Technology: Trondheim, Norway, 2017; p. 123.
37. Larsen, R. *Gamified Mobile Application to Raise Awareness and Support Energy Efficient Behavior*; Norwegian University of Science and Technology: Trondheim, Norway, 2018.
38. Klas, C.-P. *Expert Evaluation Methods, in User Studies for Digital Library Development*; Dobreva, M., O'Dwyer, A., Feliciati, F., Eds.; Facet Publishing: Cambridge, UK, 2012; pp. 75–84.
39. Oliveira, M.; Petersen, S.A. *The Choice of Serious Games and Gamification, in Serious Games Development and Applications*; Ma, M., Oliveira, M.F., Baalsrud Hauge, J., Eds.; Springer International Publishing: Cham, Switzerland, 2014.


**Publisher's Note:** MDPI stays neutral with regard to jurisdictional claims in published maps and institutional affiliations.

*Article*

# Methodology for Quantifying the Energy Saving Potentials Combining Building Retrofitting, Solar Thermal Energy and Geothermal Resources

**Silvia Soutullo [1,*], Emanuela Giancola [1], María Nuria Sánchez [1], José Antonio Ferrer [1], David García [2], María José Súarez [2], Jesús Ignacio Prieto [3], Elena Antuña-Yudego [4], Juan Luís Carús [4], Miguel Ángel Fernández [4] and María Romero [5]**

1 Department of Energy, CIEMAT, 28040 Madrid, Spain; emanuela.giancola@ciemat.es (E.G.); nuria.sanchez@ciemat.es (M.N.S.); ja.ferrer@ciemat.es (J.A.F.)
2 Department of Energy, University of Oviedo, 33203 Gijón, Spain; garciamdavid@uniovi.es (D.G.); suarezlmaria@uniovi.es (M.J.S.)
3 Department of Physics, University of Oviedo, 33203 Gijón, Spain; jprieto@uniovi.es
4 Division of Information and Technology, TSK, 33203 Gijón, Spain; elena.antuna@grupotsk.com (E.A.-Y.); juanluis.carus@grupotsk.com (J.L.C.); miguelangel.fernandez@grupotsk.com (M.Á.F.)
5 Department of Projects and R&D, GEOTER, 28703 San Sebastian de los Reyes, Spain; maria.romero@geoter.es
* Correspondence: silvia.soutullo@ciemat.es; Tel.: +34-913-466-305

Received: 9 October 2020; Accepted: 12 November 2020; Published: 16 November 2020

**Abstract:** New technological, societal and legislative developments are necessary to support transitions to low-carbon energy systems. The building sector is responsible for almost 36% of the global final energy and 40% of $CO_2$ emissions, so this sector has high potential to contribute to the expansion of positive energy districts. With this aim, a new digital Geographic Information System (GIS) platform has been developed to quantify the energy savings obtained through the implementation of refurbishment measures in residential buildings, including solar thermal collectors and geothermal technologies and assuming the postal district as the representative unit for the territory. Solar resources have been estimated from recently updated solar irradiation maps, whereas geothermal resources have been estimated from geological maps. Urbanistic data have been estimated from official cadastre databases. For representative buildings, the annual energy demand and savings are obtained and compared with reference buildings, both for heating and cooling. The GIS platform provides information on average results for each postal district, as well as estimates for buildings with particular parameters. The methodology has been applied to the Asturian region, an area of about 10,600 km$^2$ on the Cantabrian coast of Spain, with complex orography and scattered population, qualified as a region in energy transition. High rehabilitation potentials have been achieved for buildings constructed before the implementation of the Spanish Technical Building Code of 2006, being higher for isolated houses than for collective buildings. Some examples of results are introduced in specific localities of different climatic zones.

**Keywords:** energy modelling; building thermal performance; refurbishment potential; multivariable evaluation; GIS; TRNSYS

## 1. Introduction

The energy transformation towards a more efficient and less polluting system represents a global problem that must be addressed by all sectors of society. National, regional and local governments must work together involving all the stakeholders, public and private. The Intergovernmental Panel on Climate Change (IPCC) highlighted in its 2018 report [1] the strong impact produced by global warming and called for urgent actions. To reduce the high contribution of energy sector on greenhouse gas emissions (two thirds of the total), the IPCC has promoted an immediate increase in renewable energy use and energy efficiency.

The goal of achieving massive decarbonisation and reducing the rise in global temperatures to well below 2 °C [2] can be achieved safely, reliably and affordably by using sustainable technologies. The analysis develops by the International Renewable Energy Agency (IRENA) [3] shows that the combination of renewable technologies, energy efficiency and increased electrification could achieve 90% of the necessary reductions in energy-related emissions.

The generation of renewable energy depends on meteorological variables such as temperature, irradiation, precipitation or wind [4], being strongly influenced by the climatic trends of the coming years. The effects of climate change will have implications on the reliability and performance of the energy system [5,6]. This can affect the lifespan of energy infrastructure [7] and other factors within the value chain of the energy sector [8,9].

Due to the rapid population growth in cities and global climate change, energy consumption in buildings has increased in recent decades [10]. Cities in developed countries are thought to be the main worldwide energy consumers and the main source of pithy greenhouse gas emissions [11]. To tackle climate change [12] and fast urbanisation, challenges, adaptation and mitigation measures are required and play an important role in the essential transformation of urban areas. Energy potential, usage and capacity at district scale have turned into important factors to understand energy flows at this scale [13]. This knowledge can help to reduce most effectively the energy consumption and greenhouse gas emissions in the building sector [14]. Different urban models, methods and tools [15] help policymakers, governors and other stakeholders in urban design and planning, and promoting nearly zero energy buildings [16], but also in the cost-effective renovation of building stock. At the district level, urban energy use [17] has gained a prominent relevance in building refurbishment, being usually addressed by centralised interventions taking building synergies into consideration [18].

Multiple Geographic Information System (GIS) studies are related to urban energy analyses. Terés-Zubiaga et al. [19] developed a GIS-based methodology to identify optimal solutions at district scale, with balanced renewable energy supply and energy efficiency measures. The energy consumption and the renewable energy potential have also been assessed by Santoli et al. [20] at municipality scale. Other authors [21] evaluated affordable and sustainable urban electricity supply systems in cities, which have also been modelled and optimized in open source GIS platforms. Building Information Modelling (BIM) and GIS can be integrated in building environments to tackle multiple aspects such as urban governance, building energy management or construction projects [22]. A study developed by Marzouk et al. on this topic assessed the infrastructure requirements associated with the water consumption, the sewage capacity and the electrical supply for expanding cities [23]. Other authors integrated GIS into urban sustainability assessment systems, helping local governments to define land-use policies [24]. Krietemeyer et al. developed an interactive platform for spatiotemporal visualization of simulated building energy consumption to support climate adaptation strategies in variable energy scenarios [25].

In addition, energy consumption of the building stock has been assessed at the urban scale, identifying homogenous energy areas [26,27]. GIS technology has been applied in combination with statistical analysis to identify the variables that influence building energy consumption [28]. In addition, the effects of urban form [29] or climate change [30] on building demands have been evaluated. Praene et al. defined accurate climate zones to evaluate the thermal performance [31]. Garcia-Ballano et al. calculated energy savings in rehabilitated buildings after the improvement of the exterior building envelope [32]. In addition, the GIS techniques have been combined with multi-criteria decision-making tools to set up plus-energy buildings [33], and machine-learning technology has been combined with GIS to evaluate retrofit potential in buildings [34]. Finally, other studies evaluated the available renewable resources in the study area. Sarmiento et al. proposed a GIS model based on the use of satellite data to determine solar irradiation resources in a specific region of Argentina [35]. The energy potential of the biomass obtained from the urban greenery maintenance has been evaluated and has been considered as a low-range renewable resource [36]. More recently, a new GIS model has been developed to face heat waves by using pavement watering for urban cooling [37].

Moreover, these techniques have been used to obtain information about the boundary conditions of the urban environment. For example, Viana-Fonts et al. developed shadow cast profiles of buildings in urban areas from cadastral cartography and LiDAR altimetric data [38]. Liang et al. developed an interactive GIS tool for sky, tree and building view factor estimation from street view photographs [39].

To summarize, according to previous studies, the flexibility of GIS tools enables the performance of a wide variety of urban analysis.

In this context, this paper presents a new geographic information system platform developed to spatially analyse relevant urban features, such as building density and type, geothermal resources and solar potential production. Furthermore, potential energy savings on the existing building stock, considering all these factors, have been calculated. This platform relies upon GIS technology, is an initial approach to urban energy analysis and contributes to the reinforcement of European energy and climate policies oriented to a fundamental transformation of the energy system. Proposed cost-effective actions improve energy efficiency, promote centralised solutions based on low-carbon energy innovations and boost technologies to mitigate global warming. An initial case study has been applied to the Principality of Asturias, a region of more than 10,000 km$^2$ on the northern coast of Spain, with complex orography and scattered population. This region is characterized by a temperate climate based on the Köppen Geiger classification, with mild temperatures registered in both winter and summer and abundant rainfall throughout the year. There are different climate zones depending on the precipitation patterns, ranging from oceanic (Cfb) to Mediterranean oceanic (Csb). These climatic conditions lead to high heating loads and low cooling loads. These actions, in terms of reducing the carbon cycle for ongoing energy transition processes, guarantee a sustainable socio-economic development of the area, and incentivise a municipality scale approach to the implementation of renewable energies through efficient district systems.

## 2. Materials and Methods

To cope with the increase in the energy consumption in buildings, several projects have been developed during recent decades with the aim of helping users, local administrations and stakeholders to make the better decisions to improve and optimize the energy efficiency of the building stock [40].

In this line, the Spanish research project RehabilitaGeoSol (RGS) [41] has developed a citizen-oriented platform (RGS platform) that suggests to the parties involved the energy savings obtained through the implementation of retrofit interventions, solar thermal panels and geothermal technologies in residential buildings.

A building expert should choose between several retrofitting actions, reconciling energy, environmental, legal and social factors to achieve a good compromise solution that meets the final needs of occupants. The search for a reasonable solution can be carried out through an energy analysis of the building and through the development of various situations predefined by a building-physics expert, mainly through simulations [42]. In addition, a second approach which incorporates decision aid techniques such as multi-criteria analysis [43], multi-objective optimization [44], generally combined with simulations to help reach a final decision [45], can be considered.

The methodology used to create the retrofit databases of the RGS project has been applied to the Principality of Asturias. Up to 800 postal districts have been analysed, showing the interest of combining building retrofitting with solar and geothermal strategies.

Annual and seasonal thermal loads for each refurbishment option proposed have been calculated in order to provide the input data to the platform. The final outputs are the energy-savings reached by the implementation of retrofit measures, solar thermal panels and geothermal technologies in residential buildings.

### 2.1. Solar Resource

In the case of the Asturian region, there have been solar maps available since 2009 [46], which provide monthly and annual average global solar irradiation on a horizontal surface. These maps were created using indirect estimations of irradiation from air temperatures [47] and GIS techniques, and have been recently updated taking climatic data from 2007 to 2017 into account. The spatial resolution of the original maps is 25 × 25 m, but they have been converted to postal district level using ArcGIS software for spatial integration. Figure 1 shows the variations of annual average irradiation in different postal districts. Similar maps were generated for maximum, mean and minimum values of air temperature.

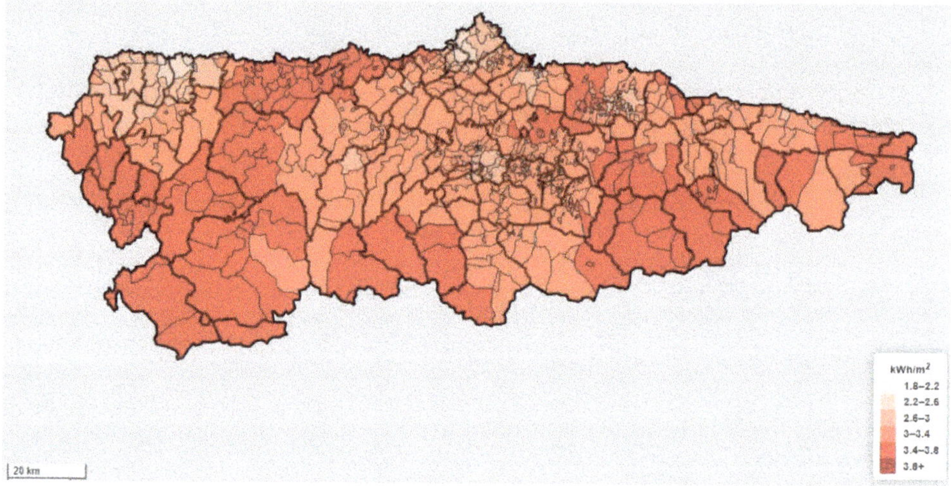


**Figure 1.** Average of daily global irradiation on a horizontal surface in Asturias.

### 2.2. Geothermal Resource

The geothermal resource is based on the National Geological Map (MAGNA), prepared by the Geological and Mining Institute of Spain (IGME) [48] between 1972 and 2003. This map consists of sheets that show geological data (geological sections, stratigraphic profiles, boreholes, etc.) with structural and hydrogeological diagrams of the national territory.

Each material is assigned some conductivity according to tabulated data by the Institute for the Diversification and Saving of Energy of Spain (IDAE) [49]. Table 1 shows some representative examples of rocks and minerals existing in the Principality of Asturias [48] and their main characteristics, among which different thermal conductivities stand out.

**Table 1.** Thermal characteristics of representative materials in Asturias.

| Type of Rock | Minimum Thermal Conductivity W/(m·K) | Average Thermal Conductivity W/(m·K) | Maximum Thermal Conductivity W/(m·K) | Volumetric Specific Heat. Capacity MJ/(m$^3$/K) |
|---|---|---|---|---|
| Basalt | 1.3 | 1.7 | 2.3 | 2.3–2.6 |
| Granite | 2.1 | 3.4 | 4.1 | 2.1–3.0 |
| Gneiss | 1.9 | 2.9 | 4.0 | 1.8–2.4 |
| Limestone | 2.5 | 2.8 | 4.0 | 2.1–2.4 |

In addition, bibliographic data are contrasted with the Enhanced Geothermal Response Test, so that the basic properties of geothermal resources have been characterized for most materials in each postal district (Figure 2). In the north-western regions, there is a high concentration of zones with values of thermal conductivity above 3 W/(m·K). In the eastern regions, the highest concentration takes place for conductivities between 2 and 3 W/(m·K). In the lower central regions, most of the zones have thermal conductivities below 2.6 W/(m·K), while in the upper central regions, most of them oscillate between 2.2 and 3 W/m K. Despite these trends, there is a great variation in conductivities throughout the Principality of Asturias.

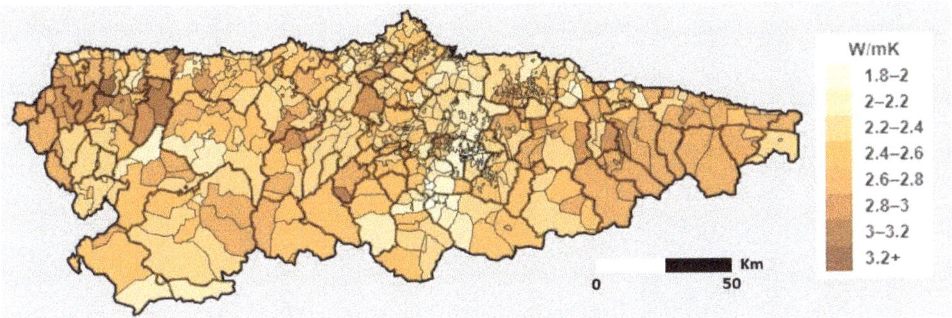


**Figure 2.** Average thermal conductivity in postal districts of the Principality of Asturias.

### 2.3. Urban Data

Official cadastre databases have been carefully handled and buildings in each postal district have been classified in the categories of Table 2. Notice that considered time periods correspond to different technical energy requirements for buildings in Spain, namely, the Spanish Basic Building Standard—NBE-CT-79 [50], the Spanish Technical Building Code—CTE [51], which entered into force at the end of 2007, and its amended version at the end of 2013.

**Table 2.** Categories used for the classification of buildings in each postal district.

| Year of Built | No. of Floors | Surface ($m^2$) | Typology |
|---|---|---|---|
| Before 1979 | 1–2 | <150 | Isolated single-family home |
| 1979–2008 | 3–5 | 150–300 | Single-family homes in closed block |
| 2009–2013 | 6–8 | 300–600 | Apartment building in open block |
| After 2014 | >8 | >600 | Apartment building in closed block |

Because of this classification, statistics have been obtained for each postal district, indicating the number of buildings of each category, which is the key information to evaluate the expected energy savings if retrofitting measures are applied.

### 2.4. Building Energy Demand

The necessary requirements to identify the most suitable refurbishment strategies to reduce the energy consumption of a building are local climates and constructive information [52]. In the first place, it is necessary to count on the availability of a representative climate data [53] of a locality. Second, it is necessary to know the characteristics of the existing building façades [54], shading devices, air ventilation [55], fenestration types and the user occupancy [56]. The study of the potential of climatic variables helps to identify the group of passive and active techniques that, when coupled, give the best result [57]. In addition to the above, the legal restrictions and the requirements of local legislation must be considered.

A dynamic simulation environment has been developed to calculate the building energy demands provided by this new digital platform [58]. This method requires an exhaustive definition of the building cases and the boundary conditions, the execution of a multi-parametric study and a pot-processing evaluation. The methodology applied is divided into three phases, as shown in Figure 3:

- Development of building models.
- Development of a dynamic simulation environment.
- Post-processing evaluation and creation of building analysis modules.

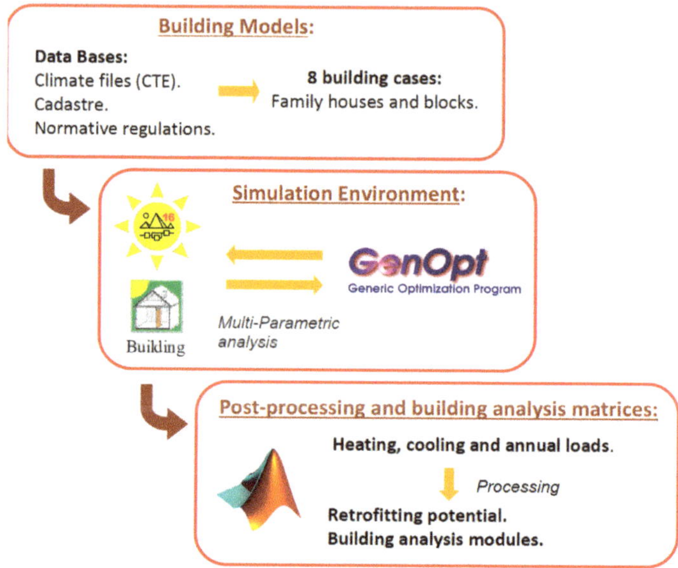


**Figure 3.** Methodology to develop building analysis modules for the online RGS platform.

2.4.1. Phase 1: Development of Building Models

The representativeness of the building stock has been selected through an input matrix that characterizes the residential building stock of the region. This information has been collected from the Spanish Technical Building Code [59], Institute for the Diversification and Saving of Energy of Spain [60], Cadastre, University of Oviedo or the Environmental Information System of the Principality of Asturias [61]. This inlet matrix is composed by the information related to the climate conditions, volume, constructive and operational characteristics.

Regarding the climate conditions, the region of Asturias is defined as temperate climate, characterized with two Köppen Geiger climatic zones: Cfb and Csb. Nevertheless, the Spanish regulation on energy savings in buildings developed a national climate classification based on seasonal severities [59]. This classification is performed as a combination of two seasonal indexes [62]: winter severity (identified by a letter) and summer severity (identified by a number). These seasonal indices are calculated as an equation that considers the influence of global solar radiation, heating degree-days and cooling degree-days [63]. Because of this climate classification, three normalized climate files are obtained for Asturias: C1, D1 and E1.

The main difference between the climate zones C1, D1 and E1 lies on the temperature and solar radiation values. The warmest zones are classified as C1 while the coldest zones are E1. Figure 4 shows the seasonal values of temperature (upper left), relative humidity (upper right), solar global radiation (lower left) and wind speed (lower right as well as the annual mean values (dark colours). Orange bars are shown for zone C1, green bars for zone D1 and blue bars for zone E1.

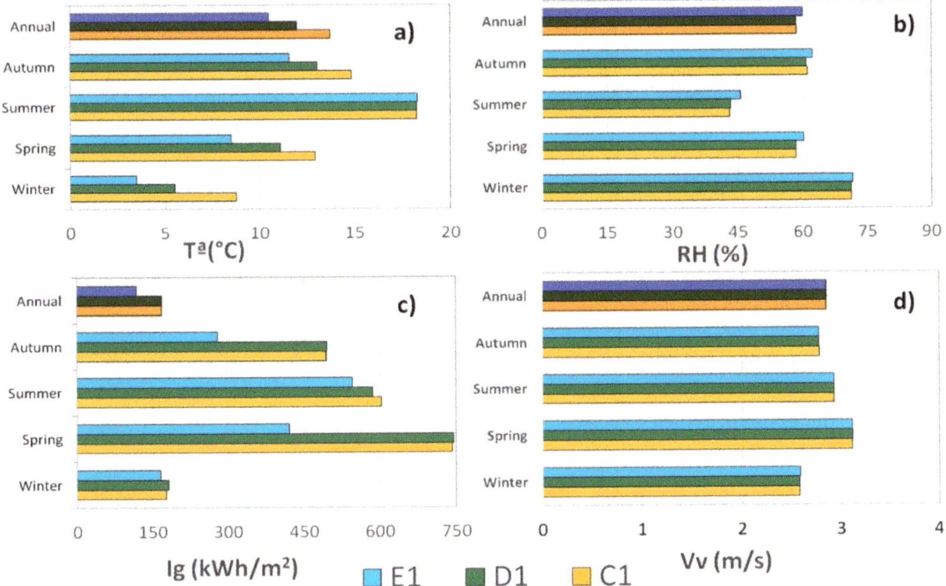


**Figure 4.** (**a**) Seasonal values of temperature, (**b**) relative humidity, (**c**) solar global radiation and (**d**) wind speed for the climate zones C1 (orange bars), D1 (green bars) and E1 (blue bars). Annual values are shown with dark colours.

Monthly mean temperature ranges from 10.5 °C (zone E1) to 13.7 °C (zone C1). The highest deviations of temperature are registered in winter, being practically non-existent during the summer. Monthly mean solar global radiation ranges from 1413 kWh/m$^2$ (zone E1) to 2007 kWh/m$^2$ and 2017 kWh/m$^2$ (zones D1 and C1, respectively). The highest values of solar global radiation are obtained for the climate zones C1 and D1, reaching greater deviations in spring and autumn and minimal in winter. The seasonal values of relative humidity and wind speed are quite similar for the three studied zones, with the maximum in spring and the minimum in winter. Monthly mean relative humidity ranges from 58.3% (zone D1) to 59.9% (zone E1). Finally, monthly mean wind speed ranges from 2.8 m/s (zone D1) to 2.9 m/s (zones C1 and E1).

As constructive characteristics, eight residential models that are representative of the Principality of Asturias have been selected based on the inlet dataset matrix. These models combine the type of houses with different configurations for the boundary conditions. Two types of residential houses have been modelled:

- Single family houses: two-storey house with 100 m$^2$ of floor area and a height between floors of 3 m (Case 1 and 2).
- Blocks of houses: four (Case 3 and 4), seven (Case 5 and 6) and ten (Case 7 and 8) square-floor plants with 3 m height between floors.

Two configurations have been modelled to consider the boundary conditions: isolated houses (odd cases), with four façades in contact with the ambient conditions; and semi-detached houses (pair cases), with two façades in contact with the ambient conditions.

Four constructive characteristics have been defined based on the Spanish regulations: before 1979, 1979–2008, 2009–2013 and after 2014. Table 3 provides the limit values of the global heat transfer coefficients for the constructive elements of the building envelope.

**Table 3.** Limit values of the overall heat transfer coefficients U ($W/m^2K$) for the building envelopes in the climatic zones.

| Climate Zone | Constructive Elements | $U_{limit}$ Before 1979 | $U_{limit}$ 1979–2008 | $U_{limit}$ 2009–2013 | $U_{limit}$ After 2014 |
|---|---|---|---|---|---|
| C1 | Roof | 2.17 | 1.20 | 0.41 | 0.23 |
| | External Wall | 2.38 | 1.60 | 0.73 | 0.29 |
| | Ground | 1.00 | 1.00 | 0.73 | 0.29 |
| | Internal Wall | 2.25 | 1.62 | 0.73 | 0.73 |
| | Glazing (g) | 5.73 (0.82) | 3.25 (0.76) | 1.54 (0.65) | 0.97 (0.61) |
| | Frame | 5.7 | 4.0 | 2.2 | 2.2 |
| D1 | Roof | 2.17 | 1.20 | 0.38 | 0.22 |
| | External Wall | 2.38 | 1.60 | 0.66 | 0.27 |
| | Ground | 1.00 | 1.00 | 0.66 | 0.27 |
| | Internal Wall | 2.25 | 1.62 | 0.66 | 0.27 |
| | Glazing (g) | 5.73 (0.82) | 3.25 (0.76) | 1.54 (0.65) | 0.97 (0.61) |
| | Frame | 5.7 | 4.0 | 2.2 | 2.2 |
| E1 | Roof | 2.17 | 1.20 | 0.35 | 0.19 |
| | External Wall | 2.38 | 1.60 | 0.57 | 0.25 |
| | Ground | 1.00 | 1.00 | 0.57 | 0.25 |
| | Internal Wall | 2.25 | 1.62 | 0.57 | 0.25 |
| | Glazing (g) | 5.73 (0.82) | 3.25 (0.76) | 1.54 (0.65) | 0.97 (0.61) |
| | Frame | 5.7 | 4.0 | 2.2 | 2.2 |

Opaque elements (roof, walls, ground) and transparent elements (glass and frames) have been defined for each climate zone and each normative value. The glass elements are characterized by the heat transfer coefficient ($U_{limit}$) and the solar heat gain coefficient (g).

In order to complete the building models, occupancy, lighting, equipment and air renovations gains are defined using the Spanish tools for the Energy Certification of Buildings as the reference database [64].

Two periods of thermal conditioning are established, summer (June–September) and winter (October–May), each one with its respective temperature set point. Two types of annual air renovation are modelled: infiltration and ventilation. The infiltration values are defined constant, depending on the type of house and the year of construction:

- Regulations before 2008: 0.8 ren/h.
- Regulations after 2008:
    - Single-family house: 0.3 ren/h.
    - Blocks of houses: 0.24 ren/h.

The ventilation rates depend on the occupation:

- No occupancy: 0.2 ren/h.
- Occupancy: 1.2 ren/h.

### 2.4.2. Phase 2: Development of a Dynamic Simulation Environment

A dynamic simulation environment has been developed to calculate the energy demands of the building cases. This environment consists on the coupling between the dynamic simulation program TRNSYS [65] and the parameterization program GenOpt [66]. The combination of these programs automates the execution of the simulation batteries; generating successive building models for each studied option. TRNSYS has been used as the engine of the simulation environment. GenOpt identifies the studied variables of the building cases and runs a series of simulations, modifying only one of them. The outputs provided by this environment are heating, cooling and annual loads and have been used to feed the building analysis tools of the online platform.

A multi-parametric analysis has been done by modifying some studied variables of the eight representative building cases.

- Climate zones: C1, D1 and E1.
- Year of construction: before 1979, 1979–2008, 2008–2013 and after 2014.
- Type of window: before 1979, 1979–2008, 2008–2013 and after 2014.
- Number of plants for blocks: 4, 7 and 10.
- Surface area for blocks: 200, 400 and 800 $m^2$.
- Percentage of shading received on windows for the main façades during summer: 0, 25, 50, 75 and 100%.

#### 2.4.3. Phase 3: Post-Processing and Creation of the Building Analysis Matrices

A post-processing evaluation of the output variables provided by the simulation batteries has been done with the program Matlab to create the databases that feed the online building modules. Heating, cooling and annual loads have been assessed to calculate the retrofitting potentials reached by each building configuration. These potentials are calculated with respect to the minimum value of the annual thermal demand, which corresponds to the normative values after 2014.

Two different analysis building modules have been developed. The first module provides the maximum retrofitting potential reached in each postal district of the Principality of Asturias. The second module provides a customized retrofitting potential for each building configuration defined by the user. Once the two databases have been created, they are represented in the GIS-based web application viewer through the Rehabilitation Potential tabs.

### *2.5. Solar Savings*

The expected energy savings caused by a domestic solar water heating (SWH) system have been estimated by means of the f-Chart method [67], considering that each single house or flat is provided with a 2 $m^2$ flat solar collector tilted 25° towards the South. The behaviour of the collector has been modelled assuming a typical performance in accordance with the following characteristic line:

$$\eta = 0.78 - 4\frac{T_e - T_{amb}}{I_g}, \tag{1}$$

where $\eta$ is the collector efficiency, $T_e$ is the fluid temperature at the inlet of the collector, $T_{amb}$ is the ambient temperature and $I_g$ is the global solar irradiance on the collector surface.

To calculate the demand in each home, a daily consumption of 120 L of domestic hot water at 60 °C has been considered.

The cold water temperature of the supply network is corrected for each postal district based on the altitude difference with the province capital, as proposed by the Spanish Technical Building Code:

$$T_r = T_{rC} - 0.005(A - A_C), \tag{2}$$

where $T_r$ is the water temperature in the supply network of the postal district, $T_{rC}$ is the water temperature in the supply network of the province capital, $A$ is the average altitude of the postal district and $A_C$ is the altitude of the province capital.

For the climatic variables, the annual average values of irradiation and air temperatures in the postal district have been used. The annual averages of irradiation on inclined collectors are calculated from the values on a horizontal surface using the following approximation, which is justified by series of measurements at the Oviedo- State Meteorological Agency of Spain (AEMET) meteorological station:

$$G(0, 25°) \approx 1.2G(0). \tag{3}$$

where $G(0, 25°)$ is the global solar irradiation on a surface tilted 25° towards the South, and $G(0)$ is the value on a horizontal surface.

The mass flow rate of 0.028 kg/s is assumed for the fluid circulating through the collector, while the installation is completed with an accumulator with a capacity of 150 litres and an internal heat exchanger with an efficiency of 90%.

## 2.6. Geothermal Savings

The geological data under Section 2.1 are used to estimate the savings derived from the geothermal resource, for a 100 $m^2$ rehabilitated single-family building, based on the building typologies and the following input variables:

- Building construction year.
- Floor surface area.
- Current fuel.
- Age of the current heating equipment.

The length of the required geothermal ground-heat exchanger is calculated as a previous step to perform a study of energy savings, $CO_2$ emissions and economic advantages compared to the current heating system, by means of the Equation (4) [49]:

$$L = \frac{Q \cdot \frac{COP-1}{COP} \cdot (R_P + R_S \cdot F)}{T_L - T_{MIN}} \tag{4}$$

where $L$ is the length of the vertical exchanger, $Q$ is the heating power of the building, COP is the equipment efficiency (Coefficient Of Performance), $R_P$ is the thermal resistance per unit length of the buried exchanger pipe, $R_S$ is the thermal resistance per unit length of the ground, $F$ is the utilization factor, $T_L$ is the ground temperature and $T_{MIN}$ is the minimum temperature.

The energy demand per surface unit of each building has been calculated keeping in mind the envelope thermal characteristics and the building typology, supported by data from the project itself. The proposed system, based on heat pump technology, is powered by geothermal energy and auxiliary electrical energy. Assuming a typical heat pump COP of 4.9, necessary electrical energy and free geothermal energy obtained from the subsoil are calculated for each building.

Geothermal systems are compared with conventional natural gas or gas oil systems in terms of $CO_2$ emissions. It takes the different performances into account according to the age of the current heating equipment and $CO_2$ emissions factors [68] (Table 4). For the final emissions calculation, these factors are multiplied by the energy consumed.

**Table 4.** $CO_2$ Emissions factors by power source.

| Power Source | $CO_2$ Emissions Factors (kg $CO_2$/kWh) | Cost—Taxes Included (€/kWh) |
|---|---|---|
| Conventional electricity | 0.331 | 0.2383 |
| Gas Oil | 0.311 | 0.0727 |
| Natural gas | 0.252 | 0.0665 |

The geothermal system is also compared in terms of energy costs. It considers the average cost in EUR per kWh of electricity [69], gas oil [70] and natural gas [69], which is also multiplied by the energy consumed (Table 4).

In this way, the results of energy, emissions and cost savings are achieved for different buildings under input parameters.

## 3. Results

A digital platform has been developed to collect all the results obtained in relation to the different energy resources studied as well as to the multi-parametric analysis of the energy performance of buildings. The main purpose of this platform is to provide end-users with the results of the energy savings assembled potential according to the interaction of the different technologies and available energy resources. This platform has been built upon GIS technologies and it is composed of two core modules which have been designed taking user-friendly requirements and an interactive visualization into account. The first module is the viewer, which is intended to provide information about the maximum potential of the energy savings achieved for building renovation in the Principality of Asturias. On the other hand, the second module of the platform makes it possible to obtain a rehabilitation potential for each customized building model evaluated. In this module, users can modify the input building conditions, adjusting them to their particular characteristics. The input variables that can be modified in a personalized way are: local climate zone, type of building, year of construction, number of floors, floor area, type of windows and percentage of summer shading on the north, south, east and west façade individually. The outputs values provided by this second module are the percentage of annual heating, cooling and total retrofitting potentials.

Figure 5 shows the main screen of the viewer module that is seen when entering the platform. Two different spaces can be visualized: a map on the right side and a space reserved for a Layer Tree on the left one. The information is shown depicted on a map of the Principality of Asturias, which is divided into postal districts. To plot the information, an enabling/disabling system is used, which can be found in the Layer Tree. When any of these layers is enabled, the associated information is depicted according to a colour scale, affecting every single postal district separately. Additional information can be obtained when the cursor is hovered over each of them.

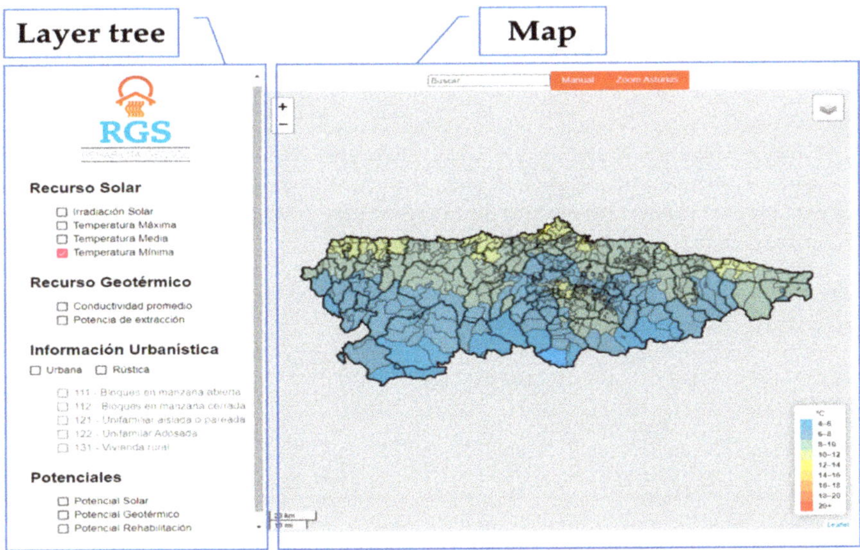


**Figure 5.** Screenshot of the platform where two areas can be distinguished: layer tree and map.

Within the different subsets of information layers belonging to the Layer Tree, a specific group called "Potentials" can be found. The different layers that make the aforementioned group provide access to the second module of the platform, the case study simulation module. This module can be accessed in the same way as the additional information pop-ups that appear in the viewer when hovering the cursor over each postal code. Once inside, there is a series of drop-down menus that allow the selection of discrete values for the definition of the specific case to be simulated. The outputs of each simulation are displayed to the user by means of numerical results or graphical representations after the selection of the relevant data.

The technologies employed in the development of the platform are open-source libraries like Leaflet, Angular and Highcharts. For the time being, the platform is only available in Spanish.

### 3.1. Solar Thermal Potential

Potential savings in domestic SWH consumption have been computed according to Section 2.5 for every postal district and represented in Figure 6. Under the assumed conditions, solar savings over 30%, which is the minimum required by the Spanish regulation for a single family house, are observed for most of the region. The software also provides the possibility of varying the number and type of collectors, as well as considering factors of loss due to orientation, tilt and shading.

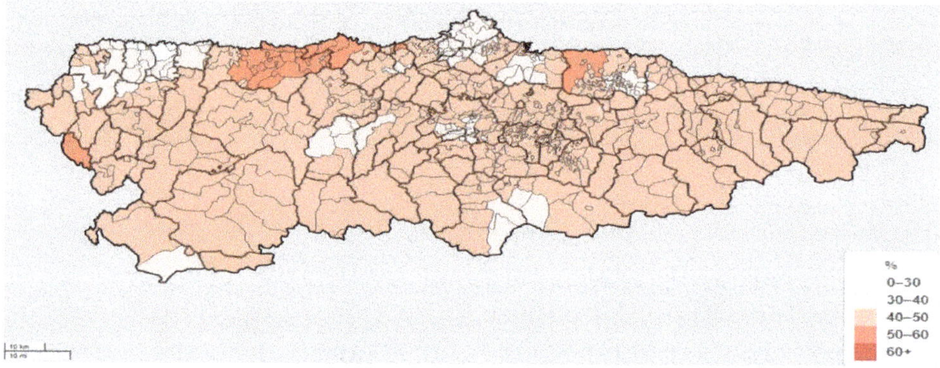


**Figure 6.** Potential savings in domestic hot water energy consumption.

Although the climatic variation within Asturias is covered with the three climatic zones, C1, D1 and E1, the new platform allows the evaluation of the general performance of solar installations, showing relevant differences in the same climatic zone. Table 5 shows the potential energy savings during a year, the annual mean efficiency of the installation for different postal districts and the yearly $CO_2$ savings, assuming a gas boiler as auxiliary energy source.

**Table 5.** Examples of variations of annual installation efficiency, solar and $CO_2$ savings for different climate zones and postal districts.

| Climatic Zone | Municipality | Postal District | Solar Savings (%) | Efficiency (%) | $CO_2$ Savings (kg $CO_2$/Year) |
|---|---|---|---|---|---|
| C1 | Gijón | 33208 | 36.6 | 45.2 | 234.3 |
| | Avilés | 33401 | 34.6 | 45.0 | 222.2 |
| | Soto del Barco | 33125 | 51.3 | 46.9 | 328.3 |
| | Coaña | 33710 | 29.6 | 43.2 | 197.3 |
| D1 | Oviedo | 33006 | 39.6 | 46.7 | 260.6 |
| | Castrillón | 33457 | 48.7 | 47.1 | 313.7 |
| | Cudillero | 33156 | 57.4 | 47.5 | 371.3 |
| | Coaña | 33716 | 26.8 | 42.3 | 172.5 |
| E1 | Cangas del Narcea | 33800 | 49.8 | 48.7 | 338.1 |
| | Grado | 33826 | 40.1 | 48.6 | 277.0 |
| | Aller | 33676 | 38.1 | 50.3 | 275.7 |

It is observed that the annual efficiency is almost constant in all climatic zones studied, while the energy savings can have great variation, particularly in zones C1 and D1. In the case of zone C1, the solar energy savings vary from 29.6% in Coaña 33710 to 51.3% in Soto del Barco 33125. A similar scenario is obtained for the D1 climatic zone, where the solar savings range from 26.8% in Coaña 33716 to 57.4% in Cudillero 33156, which are also the minimum and maximum in all the analysed area. Similar results can be observed in other cases.

### *3.2. Geothermal Potential*

The digital platform allows the calculation of the geothermal potential for each building, mainly based on the building location and the thermal conductivity of the soil. Additional input parameters for each case are: year of construction, surface area, current fuel and age of the boiler. Based on these variables, the tool determines the power extraction rate, the necessary length of the geothermal exchanger and the economic and environmental savings that the geothermal system would have in comparison to a conventional energy system.

#### 3.2.1. Geothermal Exchanger Length Depending on the Thermal Conductivity and Year of Construction

It is considered that the total heating demand is covered by geothermal heat pump. Based on this, the necessary drilling length is calculated in each case. The data of thermal conductivity in different regions are essential in order to know the drilling depth needed. There is a great impact of conductivity in these drilling meters needed to meet the energy requirements. Carrying out optimal sizing of the geothermal exchanger is essential, because geothermal drilling costs represent a high percentage of the initial investment.

As an example of the high influence of thermal conductivity on the length of the geothermal exchanger, a comparison of the same type of building is made in three locations: Gijón, Oviedo and Cangas del Narcea [48]. The type of building chosen is 100 m$^2$ and year of construction 1980–2008. The different lengths of exchanger are shown in the following figures for the locations of Gijón (Figure 7), Oviedo (Figure 8) and Cangas del Narcea (Figure 9), respectively.

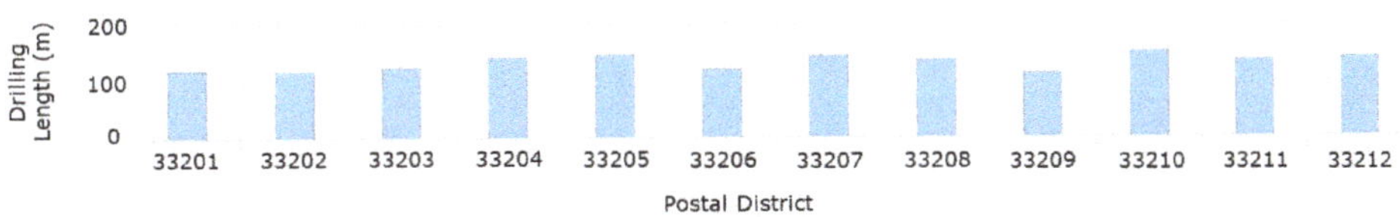


**Figure 7.** 100 m$^2$ building in Gijón. Drilling length required for each postal district.

**Figure 8.** 100 m$^2$ building in Oviedo. Drilling length required for each postal district.

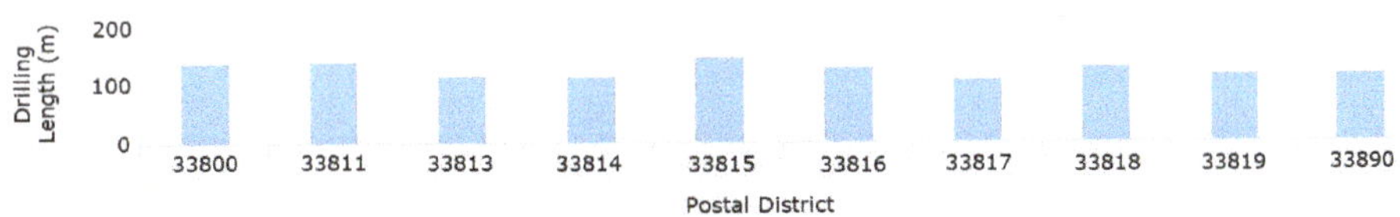


**Figure 9.** 100 m$^2$ building in Cangas del Narcea. Drilling length required for each postal district.

Keeping all other variables equal, in the postal district 33817 in Cangas del Narcea (Figure 9), 103 drilling meters are needed while in the postal district 33815, 140 m are needed. As both postal districts are in the same location, there is an increase of 35.92% affecting the cost of geothermal drilling. The larger the floor area, the more substantial the differences will be in the same locations.

The year of construction also affects geothermal sizing, as the building demand is different. In a building with a 100 $m^2$ floor plant placed in the postal district 33191 of Oviedo (Figure 8), other parameters being equal, 140 drilling meters are needed in the year of construction 1800–1979 whereas in the year of construction 2014–2016 only 97 m are needed. It is clear that building regulations have made it possible to reduce the rates of thermal demand. In this case, this is 44.33% more favourable, which undoubtedly has an impact on investment in the geothermal system.

### 3.2.2. Energy and $CO_2$ Emission Analysis

The heat pump is sized from the input data to cover all the heating building demands. It works throughout the year and the temperature of soil remains constant regardless of the outside conditions. The energy savings produced after replacing the current conventional heating system [58–60] by the geothermal heat pump system are analysed for the three previous locations. Three cases with different input values have been analysed:

- Case 1 (Figure 10): 100 $m^2$ building in Gijón (postal district 33697) constructed between 2009 and 2013, replacing a gas boiler with less than 15 years old. The geothermal system provides the total thermal energy required. In this case, it is a total of 29,334 kWh, of which 5987 kWh of electricity are consumed by the heat pump and 23,351 thermal kWh are obtained from the land, thus being free (79.59% of the total energy). The current gas boiler emits 8215 kg of $CO_2$ annually, while the proposed geothermal system would emit 1980 kg of $CO_2$. This represents an environmental saving of 6230 kg of $CO_2$.
- Case 2 (Figure 11): 200 $m^2$ building in Oviedo (postal district 33191) built between 1980 and 2008, replacing a gas boiler that is less than 15 years old. In this case, 36,079 thermal kWh are obtained free of charge from the land, which represents an economic saving of about EUR 1446. There are environmental savings of 9630 kg of $CO_2$ annually.
- Case 3 (Figure 12): 400 $m^2$ building in Cangas del Narcea (postal district 33815) built between 1800 and 1979, replacing a gas oil boiler that is more than 15 years old. Due to the age of the building and the larger floor area, the savings are higher than in cases 1 and 2. A total of 111,788 kWh are obtained free of charge per year and 47,990 kg of $CO_2$ are saved annually compared to the conventional system.

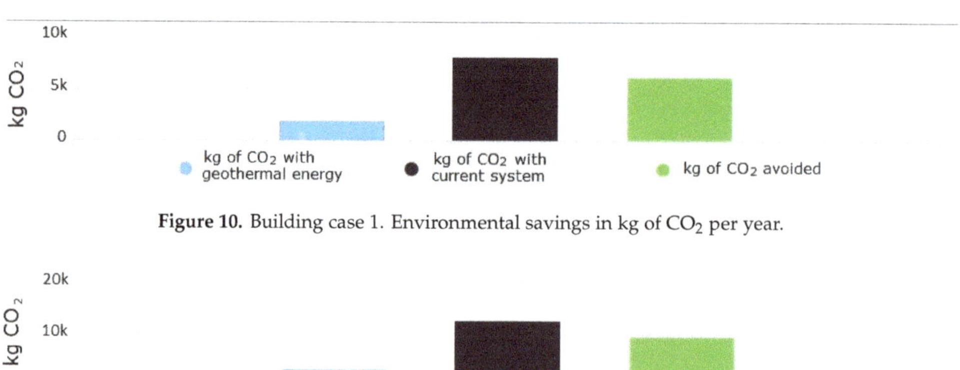


**Figure 10.** Building case 1. Environmental savings in kg of $CO_2$ per year.

**Figure 11.** Building case 2. Environmental savings in kg of $CO_2$ per year.

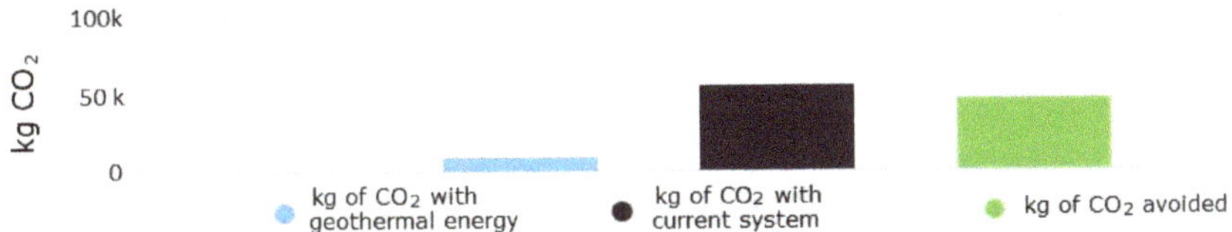


**Figure 12.** Building case 3. Environmental savings in kg of $CO_2$ per year.

The implementation of the geothermal system in cases 1 and 2 would avoid the emission of about 75% of $CO_2$ emissions, while in case 3 it is almost 84%. These data make the geothermal heating system a renewable system with special environmental interest, more accentuated in large and old buildings. In economic terms, cases 1 and 2 mean a saving of between EUR 1000 and 1500. In the case of buildings with old systems and high energy consumption, there are greater savings. In the case 3, there is about EUR 12,000 in annual savings compared to the previous gas oil system. Profitability analysis can be carried out, comparing drilling and heat pump systems versus the conventional system. In this case, with a 12 kW heat pump, it is estimated that the return on investment period is 6–7 years with an Internal Rate of Return (IRR) of around 17–18%. From an economic point of view, geothermal energy provides greater profitability than solar thermal energy, because the energy demand for SWH applications is relatively low in residential buildings, even meeting the requirements of regulations to reduce energy consumption and $CO_2$ emissions.

### 3.3. Retrofitting Potential

The mild climatic conditions registered in all regions of the Principality of Asturias have led to high percentages of heating demands in comparison with the cooling demands throughout the year. The annual heat contribution ranges, both for single-family houses and for blocks, between 71 and 93% for isolated configurations and between 83 and 95% for semi-detached configurations. The range of variability is due to climate conditions and the year of construction of the building. Figure 13 shows the heating load contribution to the annual total loads reached by the eight representative building cases (Cs1–Cs8) for the climate zones C1, D1 and E1 and for the four Spanish construction regulations. In this figure, the floor area considered for blocks is 400 $m^2$.

The heating load contributions are higher for normative regulations before 2008 and lower for normative regulations after 2008, with different percentages depending on the climate zone.

- For Spanish construction regulations before 2008 (red and orange lines):
  - Zone C1: 88 to 93%.
  - Zone D1: 90 to 94%.
  - Zone E1: 91 to 95%.
- For Spanish construction regulations after 2008 (green and blue lines):
  - Zone C1: 71 to 88%.
  - Zone D1: 74 to 90%.
  - Zone E1: 77 to 91%.

These percentages are obtained for single-family houses with 100 $m^2$ of floor area and blocks with 400 $m^2$ of floor area. In the case of blocks, the heating contributions to the annual thermal loads with floor areas of 200 $m^2$ are slightly lower while the heating contributions are slightly higher for floor areas of 800 $m^2$.

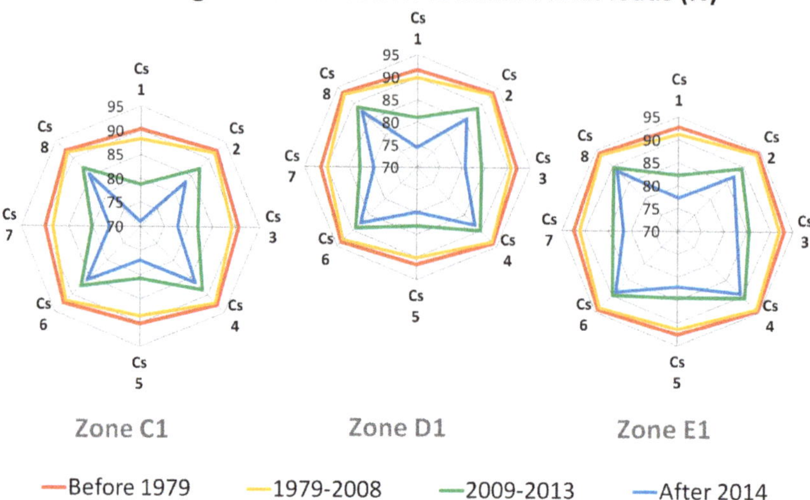


**Figure 13.** Heating load contribution to the annual total loads achieved by the eight representative building cases for the three climate zones and for the four Spanish construction regulations.

Two retrofitting potentials have been calculated superimposing the statistical study of the building stock of Asturias with the building thermal loads obtained by the simulation models. Module 1 represents the maximum retrofitting potential achieved in each studied zone of Asturias while module 2 represents the potential achieved by a specific building defined by the user.

### 3.3.1. Module 1: Maximum Retrofitting Potential

Module 1 represents the maximum retrofitting potential achieved in all the regions of the Principality of Asturias (Figure 14). These potentials, obtained with respect to the normative requirements after 2014, depend on climate conditions and the year of construction of the building. Analysing all the regions, the highest retrofitting potentials are reached in the western area of Asturias. This is due to the high percentage of single-family houses constructed before 2008, giving rise to buildings with poor energy requirements.

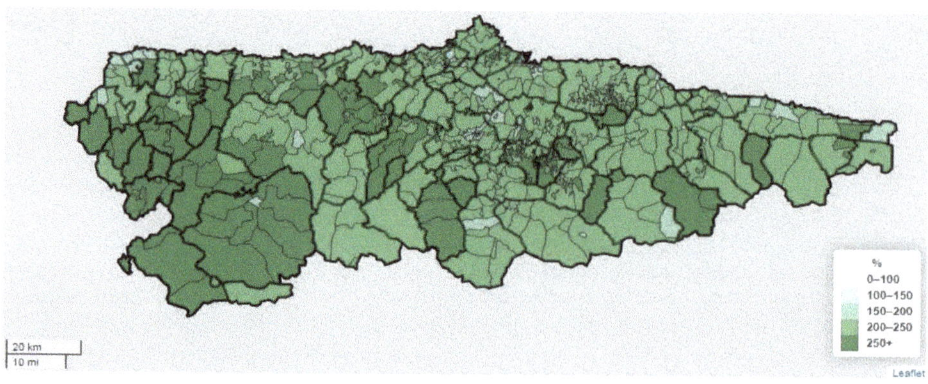


**Figure 14.** Maximum retrofitting potential shown by module 1.

With regard to the building typology, higher retrofitting percentages are achieved in single-family houses in comparison with blocks. Isolated configurations obtained higher energy potentials than semi-detached houses. There is a huge number of residential houses constructed with old construction regulations (before 2008), reaching the highest percentages of retrofitting potential.

For isolated configurations, the highest energy savings are obtained for C1 climate conditions while the lowest potentials are obtained for E1 climate conditions. For semi-detached configurations and normative requirements before 2008, the three studied climate zones reach similar energy saving potentials. For semi-detached configurations and normative requirements after 2008, the highest potentials are reached for zones C1 while the lowest potentials are reached for zones E1.

### 3.3.2. Module 2: Customize Retrofitting Potential

Module 2 supplies an estimation of customized retrofitting potentials based on the inlet variables defined by users. The use of module 2 has a great potential to quantify the energy savings achieved with the refurbishment of customized residential houses in specific regions. These potentials are obtained with respect to the minimal case that corresponds to one of the representative buildings modelled with the normative requirements after 2014.

As an example of use, 48 cases have been calculated for three municipalities of the Principality of Asturias (one for each climate zone): Gijon (zone C1), Oviedo (zone D1) and Cangas del Narcea (zone E1), as can be seen in Figure 15. Total annual retrofitting potentials are obtained for each location using module 2.

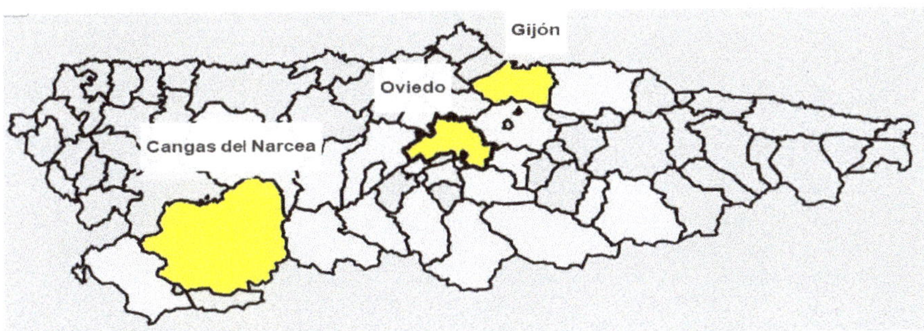


**Figure 15.** Three selected municipalities (Gijon, Oviedo and Cangas del Narcea) to assess the customized retrofitting potential with module 2.

The 48 cases analyzed in the three selected cities are summarized in Table 6 and cover:

- Two types of residential house: single-family and four-storey block. These typologies are the most usual in these regions.
- Two configurations for the boundary conditions: isolated and semi-detached.
- Four Spanish regulations to fix the construction requirements: before 1979, 1979–2008, 2009–2013 and after 2014.
- Three percentages of summer shadings over the main façades: 100, 50 and 0%. These percentages vary at the same time in all the façades.

The retrofitting potentials obtained for the 48 cases proposed in Gijon, Oviedo and Cangas del Narcea are shown in Figures 16–18, respectively. The $x$-axis represents the retrofitting potential while the $y$-axis represents the four Spanish regulations and the three summer shadings. For each construction regulation, there are a group of three summer shadings over the windows: 100% (S 100%) 50% (S 50%) and 0% (S 0%). To calculate the potential savings reached for each studied option, the constructive regulation defined after 2014 for buildings with summer shading factor of 100% over the external façades has been selected as a reference case. For this reason, all the cases 10 give a null value of retrofitting potential.

**Table 6.** Studied with module 2 for the selected places: Gijon, Oviedo and Cangas del Narcea.

| Type of House | Boundary Conditions | Normative | 100%$_t$ Summer Shading | 50%$_t$ Summer Shading | 0%$_t$ Summer Shading |
|---|---|---|---|---|---|
| Single-family | Isolated | Before 1979 | Case 1 | Case 2 | Case 3 |
| | | 1979–2008 | Case 4 | Case 5 | Case 6 |
| | | 2009–2013 | Case 7 | Case 8 | Case 9 |
| | | After 2014 | Case 10 | Case 11 | Case 12 |
| | Semi-detached | Before 1979 | Case 13 | Case 14 | Case 15 |
| | | 1979–2008 | Case 16 | Case 17 | Case 18 |
| | | 2009–2013 | Case 19 | Case 20 | Case 21 |
| | | After 2014 | Case 22 | Case 23 | Case 24 |
| Four-storey block (400 m$^2$) | Isolated | Before 1979 | Case 25 | Case 26 | Case 27 |
| | | 1979–2008 | Case 28 | Case 29 | Case 30 |
| | | 2009–2013 | Case 31 | Case 32 | Case 33 |
| | | After 2014 | Case 34 | Case 35 | Case 36 |
| | Semi-detached | Before 1979 | Case 37 | Case 38 | Case 39 |
| | | 1979–2008 | Case 40 | Case 41 | Case 42 |
| | | 2009–2013 | Case 43 | Case 44 | Case 45 |
| | | After 2014 | Case 46 | Case 47 | Case 48 |

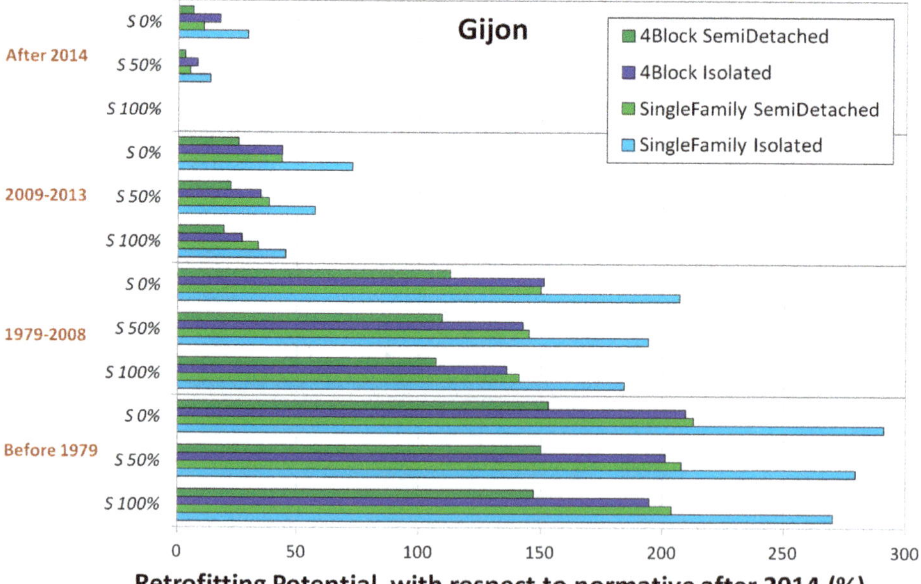


**Figure 16.** Annual retrofitting potential with respect to normative requirements after 2014 achieved in Gijon for the 48 cases proposed. Single-family houses and a four-storey block with an area of 400 m$^2$ are assessed for isolated configuration and semi-detached configurations.

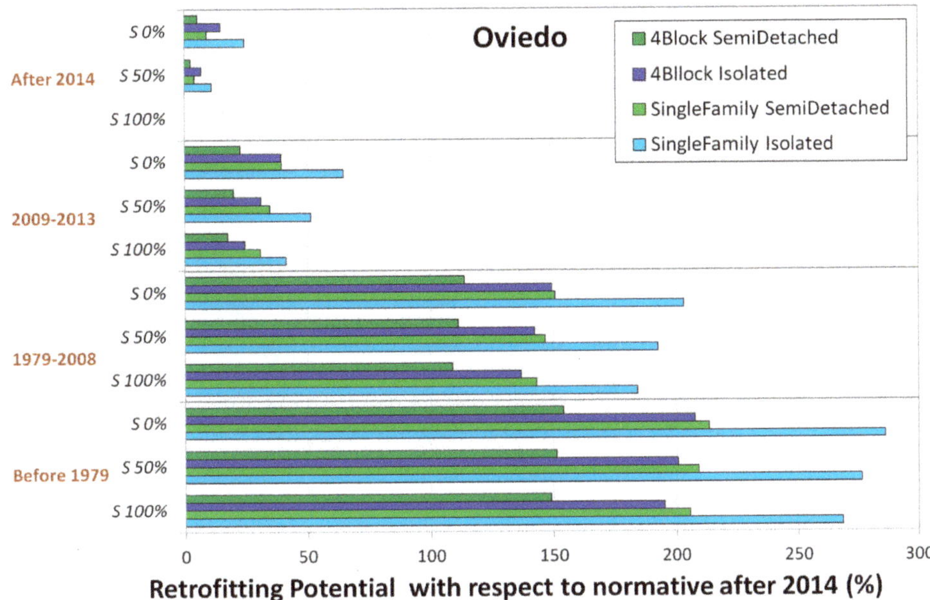


**Figure 17.** Annual retrofitting potential with respect to normative requirements after 2014 achieved in Oviedo for the 48 cases proposed. Single-family houses and a four-storey block with an area of 400 $m^2$ are assessed for isolated configuration and semi-detached configurations.

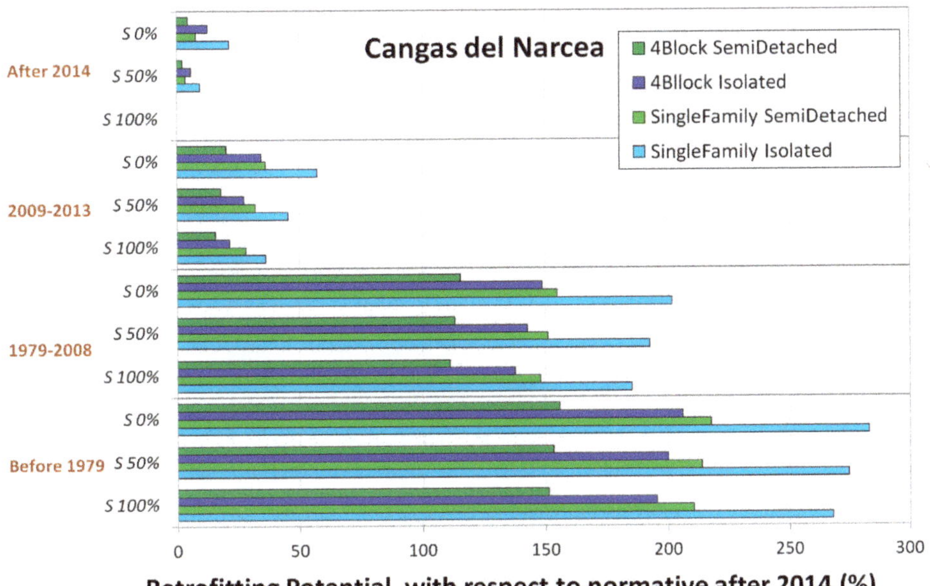


**Figure 18.** Annual retrofitting potential with respect to normative requirements after 2014 achieved in Cangas del Narcea for the 48 cases proposed. Single-family houses and a four-storey block with an area of 400 $m^2$ are assessed for isolated configuration and semi-detached configurations.

For isolated configurations, the deviation between the climate zones is very similar, being slightly higher when comparing Gijon and Oviedo. For semi-detached configurations, two tendencies have been obtained. If the construction requirements are softer (regulations before 2008), higher deviations are reached when comparing Oviedo and Cangas del Narcea. If the building energy requirements are stricter (regulations after 2008), slightly higher deviations are obtained when comparing Gijon and Oviedo.

As it can be seen in these figures, the annual retrofitting potentials reached in the three municipalities are quite similar.

In these three cities, the highest retrofitting potentials are achieved when there is no summer shading over the windows of the external façades, decreasing as the percentage of shading increases.

Higher retrofitting potentials are obtained for single-family houses in comparison with four-storey blocks, especially when the construction requirements are softer (before 2008). In both types of houses, higher potentials are achieved for isolated configurations, being more remarkable for construction regulations before 2008.

## 4. Discussion and Conclusions

One of the results obtained in the RehabilitaGeoSol project is the creation of a GIS platform developed to quantify the energy-savings obtained through the implementation of retrofit measures, solar thermal panels and geothermal technologies in residential buildings placed in the Principality of Asturias. This platform is composed of two different modules. Module 1 calculates, for each district of Asturias, the maximum energy-saving potential. Module 2 makes it possible to obtain a personalized potential based on the input characteristic of the platform.

The new developed GIS platform based on the interaction of different technologies and available resources is a step forward in the field of built environment towards smart energy communities, supporting the low carbon energy transition. The obtained results point to great energy mitigation potential, justifying the need for public authorities to develop policies aimed at reducing the energy demand of buildings, especially the heating loads. The generated maps could be helpful for retrofitting considerations for existing housing stock in the Principality of Asturias, a key action that is completed in combination with the use of solar thermal energy and geothermal resources towards sustainable development. These policies are aligned with current construction trends of reducing the energy use during the lifecycle of buildings and the promotion of the low carbon circular economy. Not only environmental but also economical reasons promote building retrofitting vs. demolition and new construction [71].

These building adaptation practices will also help address climate change, optimal use of energy resources, greenhouse emissions and secure energy supply in the coming years, by gradually implementing Net Zero Energy Buildings and Positive Energy Districts.

The RGS platform results show the high retrofitting potential for the oldest Spanish regulations (before 2006). Eight building models (Case 1–Case 8) have been developed with the dynamic simulation program TRNSYS to characterize the energy performance of residential buildings for three climatic zones (C1, D1, and E1), four Spanish construction regulations and different percentages of summer shading on the windows. The potential is higher in houses than in blocks, being more remarkable in single or isolated building configurations. In the case of blocks, the retrofitting potential is higher for four-storey plants in comparison with seven and ten-storey plans. If the surface area of blocks is higher, the energy-savings obtained are lower. The highest energy-saving potential is reached in warmer climates (zones C1), while the lowest is reached in colder zones (zones E1). Analysing the influence of external shadings on facades, this platform obtains higher impact on buildings constructed according to stricter regulations and in single-family structures vs. blocks. Regarding climate impact, the annual thermal load increases in colder climates due to the higher heating requirements. On the contrary, the impact of shading is greater in warmer climates where the cooling demands are higher.

Local authorities and other stakeholders have to consider the retrofit potential as well as the solar and geothermal resources to establish the priority areas of intervention. The expected energy savings caused by the solar domestic hot water system have been computed by means of the f-chart method assuming that each single house or flat is provided with a 2 $m^2$ flat solar collector tilted 25° towards the South. Results show that almost all the territory has solar savings over 30%, which is the minimum required by the Spanish regulations for a single-family house. The new tool allows evaluation of the general performance of solar installations, showing relevant differences in the same climate zone. The annual efficiency is almost constant in all climate zones studied; the energy savings show great variation, particularly in zones C1 and D1. In the case of zone C1, the solar energy savings vary from 29.6% to 51.3%. A similar scenario is obtained for the D1 climate zone, where the solar savings range from 26.8% to 57.4%. South-West postal districts of the Principality of Asturias present high values of the annual average of daily global solar irradiation on horizontal surfaces as well as high retrofitting potential. However, the highest values of thermal conductivity are located in north-west and east postal districts. These geological data obtained are used to calculate the length of the required geothermal ground-heat exchanger. Energy demand per surface of each building has been calculated considering the thermal envelope characteristics and the building typology. The proposed system, based on heat pump technology, is powered by geothermal energy and an input of electrical energy. The geothermal system is compared with conventional natural gas or gas oil systems in terms of $CO_2$ emissions. A comparison of the high influence of thermal conductivity on the length of the geothermal exchanger, for the same type of building, is made in three climatic zones (C1, D1, E1).

On the one hand, the influence of thermal conductivity on the length of the exchanger can be appreciated, requiring fewer drilling meters for higher conductivities. Furthermore, it can be seen that within the same location the hydrogeology is different, requiring different lengths of exchanger for each particular case. On the other hand, the factors that most influence energy and environmental savings are the improvement of the building envelope and the replacement of the conventional heating system by the geothermal one. Greater benefits are obtained for old buildings with large surfaces, with insulation or with old heating systems based on gas oil boilers. This has a direct effect on the initial investment and subsequent amortization.

The implementation of both renewable technologies in residential buildings reduces the $CO_2$ emissions to the atmosphere, with the geothermal pumps being more profitable than the solar thermal panels due to the percentage of annual thermal load cover by each technology.

It should be noted that this accessible GIS platform provides both the solar and geothermal sectors with relevant information on these renewable resources in the areas of possible implementation, solving one of the major obstacles in the implementation of low enthalpy geothermal energy and solar thermal energy in urbanized areas. Therefore, the exploitation of this GIS platform by different public and private organizations is forecast. Regarding the public entities, the platform is mainly destined to be of use to environmental government areas of the Principality of Asturias, as well as to local authorities, helping them with decision-making in the matter of energy efficiency strategies and setting priorities for programs linked to rehabilitation plans and energy actions. On the other hand, the possibility of commercialization to private companies is focused mainly on the technical or engineering sector. These companies could find, in this platform, valuable information about the areas to which they could direct their main efforts to promote their activity.

**Author Contributions:** Conceptualization, J.I.P. and J.A.F.; methodology, S.S., E.G., M.N.S., D.G. and M.J.S.; software, J.L.C. and M.Á.F.; renewable modelling, M.R. and D.G.; building modelling, S.S., E.G. and M.N.S.; renewable analysis, M.R., D.G. and M.J.S.; building analysis, E.G., M.N.S. and S.S.; writing—original draft preparation, S.S., E.G., M.N.S., D.G. and M.J.S.; writing—review and editing, M.R., E.A.-Y., S.S., E.G., M.N.S., J.A.F., D.G., J.I.P. and M.J.S. All authors have read and agreed to the published version of the manuscript.

**Funding:** This research has been developed in the framework of the REHABILITAGEOSOL (RTC-2016-5004-3) project. It is a multidisciplinary R&D program supported by the Spanish Ministry of Science and Innovation and co-financed by European Regional Development Funds (ERDF).

**Acknowledgments:** The authors would like to acknowledge the support given by the rest of the members and institutions participating in the REHABILITAGEOSOL (RTC-2016-5004-3) project. The computational work has been carried out using the computer facilities of the Extremadura Center for Advanced Technologies (CETA-CIEMAT).

**Conflicts of Interest:** The authors declare no conflict of interest.

## References

1. IPCC. *Global Warming of 1.5 °C. An IPCC Special Report on the Impacts of Global Warming of 1.5 °C above Pre-Industrial Levels and Related Global Greenhouse Gas Emission Pathways, in the Context of Strengthening the Global Response to the Threat of Climate Change, Sustainable Development, and Efforts to Eradicate Poverty;* Masson-Delmotte, V., Zhai, P., Pörtner, H.-O., Roberts, D., Skea, J., Shukla, P.R., Pirani, A., Moufouma-Okia, W., Péan, C., Pidcock, R., et al., Eds.; IPCC: Geneva, Switzerland, 2018.
2. Solauna, K.; Cerdá, E. Climate change impacts on renewable energy generation. A review of quantitative projections. *Renew. Sustain. Energy Rev.* **2019**, *116*, 109415. [CrossRef]
3. IRENA. Climate Change and Renewable Energy: National policies and the role of communities, cities and regions. In *Report to the G20 Climate Sustainability Working Group (CASE WG);* International Renewable Energy Agency: Abu Dhabi, UAE, 2019.
4. Contreras-Lisperguer, R.; de Cuba, K. *The Potential Impact of Climate Change on the Energy Sector in the Caribbean Region;* Organization of American States: Washington, DC, USA, 2008.
5. Cronin, J.; Anandarajah, G.; Dessens, O. Climate change impacts on the energy system: A review of trends and gaps. *Clim. Chang.* **2018**, *151*, 79–93. [CrossRef] [PubMed]
6. Arent, D.J.; Tol, R.S.J.; Faust, E.; Hella, J.P.; Kumar, S.; Strzepek, K.M.; Tóth, F.L.; Yan, D.; Abdulla, A.; Kheshgi, H.; et al. Chapter 10 Key Economic Sectors and Services. In *Climate Change 2014 Impacts, Adaptation and Vulnerability: Part A: Global and Sectoral Aspects;* Cambridge University Press: Cambridge, UK, 2010; pp. 659–708.
7. Ebinger, J.; Vergara, W. *Climate Impacts on Energy Systems: Key Issues for Energy Sector Adaptation;* The International Bank for Reconstruction and Development/The World Bank: Washington, DC, USA, 2011.
8. Johnston, P.C. *Climate Risk and Adaptation in the Electric Power Sector;* Asian Development Bank: Manila, Philippines, 2012.
9. Schaeffer, R.; Szklo, A.S.; de Lucena, A.F.P.; Borba, B.S.M.C.; Nogueira, L.P.P.; Fleming, F.P.; Troccoli, A.; Harrison, M.; Boulahya, M.S. Energy sector vulnerability to climate change: A review. *Energy* **2012**, *38*, 1–12. [CrossRef]
10. Sánchez, M.N.; Soutullo, S.; Olmedo, R.; Bravo, D.; Castaño, S.; Jiménez, M.J. An experimental methodology to assess the climate impact on the energy performance of buildings: A ten-year evaluation in temperate and cold desert areas. *Appl. Energy* **2020**, *264*, 114730. [CrossRef]
11. Cao, X.; Dai, X.; Liu, J. Building energy-consumption status worldwide and the state-of-the-art technologies for zero-energy buildings during the past decade. *Energy Build.* **2016**, *128*, 198–213. [CrossRef]
12. Soutullo, S.; Giancola, E.; Jiménez, M.J.; Ferrer, J.A.; Sánchez, M.N. How Climate Trends Impact on the Thermal Performance of a Typical Residential Building in Madrid. *Energies* **2020**, *13*, 237. [CrossRef]
13. European Energy Research Alliance. EERA JPSC MAPPING: Positive Energy Districts (PEDs) and Neighbourhoods. Available online: https://www.eera-set.eu/eera-joint-programmes-jps/list-of-jps/smart-cities/mapping-positive-energy-districts-neighbourhoods/ (accessed on 30 September 2020).
14. Temporary Working Group of the European Strategic Energy Technology (SET)-Plan on Action 3.2 "Smart Cities and Communities". Implementation Plan, Europe to become a Global Role Model in Integrated, Innovative Solutions for the Planning, Deployment, and Replication of Positive Energy Districts. Available online: https://setis.ec.europa.eu/system/files/setplan_smartcities_implementationplan.pdf (accessed on 30 September 2020).
15. Li, C. GIS for Urban Energy Analysis. *Compr. Geogr. Inf. Syst.* **2018**, 187–195. [CrossRef]
16. Cabeza, L.F.; Chàfer, M. Technological options and strategies towards zero energy buildings contributing to climate change mitigation: A systematic review. *Energy Build.* **2020**, *219*, 110009. [CrossRef]

17. Abbasabadi, N.; Ashayeri, M. Urban energy use modeling methods and tools: A review and an outlook. *Build. Environ.* **2019**, *161*, 106270. [CrossRef]
18. Soutullo, S.; Giancola, E.; Heras, M.R. Dynamic energy assessment to analyze different refurbishment strategies of existing dwellings placed in Madrid. *Energy* **2018**, *152*, 1011–1023. [CrossRef]
19. Terés-Zubiaga, J.; Bolliger, R.; Almeida, M.G.; Barbosa, R.; Rose, J.M.; Thomsen, K.E.; Montero, E.; Briones-Llorente, R. Cost-effective building renovation at district 1 level combining energy efficiency & renewable. Methodology assessment proposed in IEA-Annex 75 and a demonstration case study. *Energy Build.* **2020**, *224*, 110280.
20. de Santoli, L.; Mancini, F.; Garcia, D.A. A GIS-based model to assess electric energy consumptions and usable renewable energy potential in Lazio region at municipality scale. *Sustain. Cities Soc.* **2019**, *46*, 101413. [CrossRef]
21. Alhamwi, A.; Medjroubi, W.; Vogt, T.; Agert, C. FlexiGIS: An open source GIS-based platform for the optimisation of flexibility options in urban energy systems. In Proceedings of the CUE2018-Applied Energy Symposium and Forum 2018: Low Carbon Cities and Urban Energy Systems, Shanghai, China, 5–7 June 2018.
22. Wang, H.; Pan, Y.; Luo, X. Integration of BIM and GIS in sustainable built environment: A review and bibliometric analysis. *Autom. Constr.* **2019**, *103*, 41–52. [CrossRef]
23. Marzouk, M.; Othman, A. Planning utility infrastructure requirements for smart cities using the integration between BIM and GIS. *Sustain. Cities Soc.* **2020**, *57*, 102120. [CrossRef]
24. Pedro, J.; Silva, C.; Duarte Pinheiro, M. Integrating GIS spatial dimension into BREEAM communities sustainability assessment to support urban planning policies, Lisbon case study. *Land Use Policy* **2019**, *83*, 424–434. [CrossRef]
25. Krietemeyer, B.; El Kontar, R. A method for integrating an UBEM with GIS for spatiotemporal visualization and analysis. In Proceedings of the 10th Annual Symposium on Simulation for Architecture and Urban Design Conference SimAUD 2019, Atlanta, GA, USA, 7–9 April 2019; Simulation Series; Volume 51, pp. 87–94.
26. Gaspari, J.; De Giglio, M.; Antonini, E.; Vodola, V. A GIS-Based Methodology for Speedy Energy Efficiency Mapping: A Case Study in Bologna. *Energies* **2020**, *13*, 2230. [CrossRef]
27. Li, C.; Song, Y.; Kaza, N.; Burghardt, R. Explaining Spatial Variations in Residential Energy Usage Intensity in Chicago: The Role of Urban Form and Geomorphometry. *J. Plan. Educ. Res.* **2019**, 1–15. [CrossRef]
28. Torabi Moghadam, S.; Toniolo, J.; Mutani, G.; Lombardi, P. A GIS-statistical approach for assessing built environment energy use at urban scale. *Sustain. Cities Soc.* **2018**, *37*, 70–84. [CrossRef]
29. Ahn, Y.; Sohn, D.W. The effect of neighbourhood-level urban form on residential building energy use: A GIS-based model using building energy benchmarking data in Seattle. *Energy Build.* **2019**, *196*, 124–133. [CrossRef]
30. Zheng, Y.; Weng, Q. Modeling the effect of climate change on building energy demand in Los Angeles county by using a GIS-based high spatial- and temporal resolution approach. *Energy* **2020**, *176*, 641–655. [CrossRef]
31. Praene, J.P.; Malet-Damour, B.; Radanielina, M.H.; Fontaine, L.; Rivière, G. GIS-based approach to identify climatic zoning: A hierarchical clustering on principal component analysis. *Build. Environ.* **2019**, *164*, 106330. [CrossRef]
32. García-Ballano, C.J.; Ruiz-Varona, A.; Casas-Villarreal, L. Parametric-based and automatized GIS application to calculate energy savings of the building envelope in rehabilitated nearly zero energy buildings (nZEB). Case study of Zaragoza, Spain. *Energy Build.* **2020**, *215*. [CrossRef]
33. Mokhtara, C.; Negrou, B.; Settou, N.; Gouareh, A.; Settou, B. Pathways to plus-energy buildings in Algeria: Design optimization method based on GIS and multi-criteria decision-making. *Energy Proc.* **2019**, *162*, 171–180. [CrossRef]
34. Re Cecconi, F.; Moretti, N.; Tagliabue, L.C. Application of artificial neutral network and geographic information system to evaluate retrofit potential in public school buildings. *Renew. Sustain. Energy Rev.* **2019**, *110*, 266–277. [CrossRef]
35. Sarmiento, N.; Belmonte, S.; Dellicompagni, P.; Franco, J.; Escalante, K.; Sarmiento, J. A solar irradiation GIS as decision support tool for the Province of Salta, Argentina. *Renew. Energy* **2018**, *132*, 68–80. [CrossRef]
36. Ferla, G.; Caputo, P.; Colaninno, N.; Morello, E. Urban greenery management and energy planning: A GIS-based potential evaluation of pruning by-products for energy application for the city of Milan. *Renew. Energy* **2020**, *160*, 185–195. [CrossRef]

37. Hendel, M.; Bobée, C.; Karam, G.; Parison, S.; Berthe, A.; Bordin, P. Developing a GIS tool for emergency urban cooling in case of heat-waves. *Urban Clim.* **2020**, *33*, 100646. [CrossRef]
38. Viana-Fons, J.D.; Gonzálvez-Maciá, J.; Payá, J. Development and validation in a 2D-GIS environment of a 3D shadow cast vector-based model on arbitrarily orientated and tilted surfaces. *Energy Build.* **2020**, *224*, 110258. [CrossRef]
39. Liang, J.; Gong, J.; Zhang, J.; Li, Y.; Wu, D.; Zhang, G. GSV2SVF-an interactive GIS tool for sky, tree and building view factor estimation from street view photographs. *Build. Environ.* **2020**, *168*, 106475. [CrossRef]
40. Soutullo, S.; Giancola, E.; Franco, J.M.; Boton, M.; Ferrer, J.A.; Heras, M.R. New simulation platform for the rehabilitation of residential building in Madrid. *Energy Proc.* **2017**, *122*, 817–822. [CrossRef]
41. RehabilitaGeoSol Project Website. Available online: http://projects.ciemat.es/web/rehabilitageosol/inicio (accessed on 30 September 2020).
42. Harish, V.S.K.V.; Kumar, A. A review on modeling and simulation of buildings energy systems. *Renew. Sustain. Energy Rev.* **2016**, *56*, 1272–1292. [CrossRef]
43. Ascione, F.; Bianco, N.; Mauro, G.M.; Napolitano, D.F.; Vanoli, G.P. A Multi-Criteria Approach to Achieve Constrained Cost-Optimal Energy Retrofits of Buildings by Mitigating Climate Change and Urban Overheating. *Climate* **2018**, *6*, 37. [CrossRef]
44. Hashempour, N.; Taherkhani, R.; Mahdikhani, M. Energy performance optimization of existing buildings: A literature review. *Sustain. Cities Soc.* **2020**, *54*, 101967. [CrossRef]
45. Castro, S.S.; López, M.J.S.; Menéndez, D.G.; Marigorta, E.B. Decision matrix methodology for retrofitting techniques of existing buildings. *J. Clean. Product.* **2019**, *240*, 118153. [CrossRef]
46. Prieto, J.I.; Martínez-García, J.C.; García, D.; Santoro, R.; Rodríguez, A. *Mapa Solar de Asturias*; Empresas Acuerdo Consorcio PSE-ARFRISOL: Gijón, Spain, 2009.
47. Prieto, J.I.; Martínez-García, J.C.; García, D. Correlation between global solar irradiation and air temperature in Asturias, Spain. *Sol. Energy* **2009**, *83*, 1076–1085. [CrossRef]
48. Instituto Geológico y Minero de España (IGME). Mapa Hidrogeológico de España y Mapa Geológico de España (MAGNA 50). Sheets 10–15, 25–32, 49–56, 74–80, 99–103. 1972–2003. Available online: http://info.igme.es/cartografiadigital/geologica/Magna50.aspx (accessed on 30 September 2020).
49. Asociación Técnica Española de Climatización y Refrigeración (ATECYR). *Guía Técnica: Diseño de Sistemas de Intercambio Geotérmico de Circuito Cerrado*; Instituto para la Diversificación y Ahorro de la Energía (IDAE), Ministry of Industry, Energy and Tourism, Spanish Government: Madrid, Spain, 2012; ISBN 978-84-96680-60-9.
50. Spanish Government. *Norma Básica de la Edificación "NBE-CT-79" Sobre Condiciones Térmicas de los Edificios*; Royal Decree 2429/1979 1979, BOE-A-1979-24866; Spanish Government: Madrid, Spain, 1979.
51. Spanish Government. *Código Técnico de la Edificación*; Royal Decree 314/2006 2006, BOE-A-2006-5515; Spanish Government: Madrid, Spain, 2006.
52. Soutullo, S.; Sánchez, M.N.; Enríquez, R.; Olmedo, R.; Jiménez, M.J. Bioclimatic vs conventional building: Experimental quantification of the thermal improvements. *Energy Proc.* **2017**, *122*, 823–828. [CrossRef]
53. Soutullo, S.; Sánchez, M.N.; Enríquez, R.; Jiménez, M.J.; Heras, M.R. Empirical estimation of the climatic representativeness in two different areas: Desert and Mediterranean climates. *Energy Proc.* **2017**, *122*, 829–834. [CrossRef]
54. Sánchez, M.N.; Giancola, E.; Blanco, E.; Soutullo, S.; Suárez, M.J. Experimental Validation of a Numerical Model of a Ventilated Façade with Horizontal and Vertical Open Joints. *Energies* **2020**, *13*, 146. [CrossRef]
55. Salcido, J.C.; Raheem, A.A.; Isaa, R.A. From simulation to monitoring: Evaluating the potential of mixed-mode ventilation (MMV) systems for integrating natural ventilation in office buildings through a comprehensive literature review. *Energy Build.* **2016**, *127*, 1008–1018. [CrossRef]
56. Díaz, J.A.; Jiménez, M.J. Experimental assessment of room occupancy patterns in an office building. Comparison of different approaches based on $CO_2$ concentrations and computer power consumption. *Appl. Energy* **2017**, *199*, 121–141. [CrossRef]
57. Tejero-González, A.; Andrés-Chicote, M.; García-Ibáñez, P.; Velasco-Gómez, E.; Rey-Martínez, F.J. Assessing the applicability of passive cooling and heating techniques through climate factors: An overview. *Renew. Sustain. Energy Rev.* **2016**, *65*, 727–742. [CrossRef]

58. Giancola, E.; Sánchez, M.N.; Ferrer, J.A.; López, H.; Soutullo, S. New Platform to Quantify the Energy-Saving Potential in Existing Residential Buildings in a Northern Spanish Region. In Proceedings of the 35th International Conference PLEA 2020: Sustainable Architecture and Urban Design Planning Post Carbon Cities, A Coruña, Spain, 1–3 September 2020.
59. Spanish Building Code Webpage. Available online: https://www.codigotecnico.org/index.php/menu-ahorro-energia.html (accessed on 30 September 2020).
60. Carrera, A.; Sisó, L.; Herena, A.; Valle, M.; Casanova, M.; González, D. *Evaluación del Potencial de Climatización con Energía Solar Térmica en Edificios*; Estudio Técnico PER 2011–2020; Instituto para la Diversificación y Ahorro de la Energí (IDAE): Madrid, Spain, 2011.
61. Environmental Information System of the Principality of Asturias Webpage. Available online: https://www.asturias.es/portal/site/medioambiente/menuitem.4691a4f57147e2c2553cbf10a6108a0c/?vgnextoid=eaddffae3867b210VgnVCM10000097030a0aRCRD&i18n.http.lang=en (accessed on 30 September 2020).
62. Peci Lopez, F.; de Adana Santiago, M.R. Sensitivity study of an opaque ventilated façade in the winter season in different climate zones in Spain. *Renew. Energy* **2015**, *75*, 524–533. [CrossRef]
63. Verichev, K.; Carpio, M. Climatic zoning for building construction in a temperate climate of Chile. *Sustain. Cities Soc.* **2018**, *40*, 352–364. [CrossRef]
64. Spanish Energy Certification of Buildings Webpage. Available online: https://energia.gob.es/desarrollo/EficienciaEnergetica/CertificacionEnergetica/DocumentosReconocidos/Paginas/documentosreconocidos.aspx (accessed on 30 September 2020).
65. TRNSYS Webpage. Available online: http://www.trnsys.com/ (accessed on 30 September 2020).
66. GENOPT Webpage. Available online: https://simulationresearch.lbl.gov/GO/ (accessed on 30 September 2020).
67. Klein, S.A.; Beckman, W.A.; Duffie, J.A. A Design Procedure for Solar Heating Systems. *Sol. Energy* **1976**, *18*, 113–127. [CrossRef]
68. Ministerio de Industria, Energía y Turismo y Ministerio de Fomento. *Documento Reconocido del Reglamento de Instalaciones Térmicas en los Edificios (RITE). Factores de Emisión de $CO_2$ y Coeficientes de Paso a Energía Primaria de Diferentes Fuentes de Energía Final Consumidas en el Sector de Edificios en España*; Spanish Goverment: Madrid, Spain, 2016.
69. Eurostat 2018 Webpage. Available online: https://ec.europa.eu/eurostat/data/database (accessed on 30 September 2020).
70. Instituto para la Diversificación y Ahorro de la Energía (IDAE). *Informe de Precios Energéticos: Combustibles y Carburantes*; Ministry for Ecological Transition, Spanish Goverment: Madrid, Spain, 2018.
71. Rodríguez, A.; Martínez, M.D.; González, A.; Ferreira, P.; Marrero, M. Building rehabilitation versus demolition and new construction: Economic and environmental assessment. *Environ. Impact Assess. Rev.* **2017**, *66*, 115–126. [CrossRef]

*Perspective*

# Sustainable Urban Areas for 2030 in a Post-COVID-19 Scenario: Focus on Innovative Research and Funding Frameworks to Boost Transition towards 100 Positive Energy Districts and 100 Climate-Neutral Cities

**Paola Clerici Maestosi [1,*], Maria Beatrice Andreucci [2] and Paolo Civiero [3]**

1 Department of Energy Technologies and Renewables TERIN-SEN Smart Energy Networks, ENEA—Italian National Agency for New Technologies, Energy and Sustainable Economic Development, Lungotevere Thaon de Revel, 00196 Rome, Italy
2 Department of Planning, Design, Technology of Architecture, Sapienza University of Rome, Piazzale Aldo Moro, 00185 Rome, Italy; mbeatrice.andreucci@uniroma1.it
3 Thermal Energy and Building Performance Group, IREC—Catalonia Institute for Energy Research, Sant Adrià del Besos, 08930 Barcelona, Spain; pciviero@irec.cat
* Correspondence: paola.clerici@enea.it

**Abstract:** Cities generate about 85% of the EU's GDP. As such, they are key players in shaping and providing technological and social innovations but also environmental impact. Thus, they must urgently engage in unprecedented systemic transformational and bold transitions towards sustainability and climate neutrality. The contribution—taking into account that the concepts of community resilience and urban transition have changed as a consequence of COVID-19—critically discusses innovative frameworks and funding opportunities that Horizon Europe will put in place to boost sustainable urban areas in Europe, driving a transition to 100 Positive Energy Districts and 100 climate-neutral cities by 2030.

**Keywords:** positive energy district; climate-neutral cities; 15-minute city; downsizing district; doughnuts; post-COVID-19 scenario for RD&I funding

**Citation:** Clerici Maestosi, P.; Andreucci, M.B.; Civiero, P. Sustainable Urban Areas for 2030 in a Post-COVID-19 Scenario: Focus on Innovative Research and Funding Frameworks to Boost Transition towards 100 Positive Energy Districts and 100 Climate-Neutral Cities. *Energies* **2021**, *14*, 216. https://doi.org/10.3390/en14010216

Received: 9 December 2020
Accepted: 29 December 2020
Published: 4 January 2021



## 1. Introduction

The discussion about sustainable development of urban areas started worldwide more than 30 years ago when the European Commission began to foster sustainable urban development through the Framework Programs FP5–FP6–FP7 (Appendix A), supporting urban areas to become Actors of Open Innovation, thus accelerating the transition to sustainable, low-carbon societies.

The H2020 Framework Program addressed Major Urban Challenges, stimulating cities to be Actors of Open Innovation and calling municipal authorities to be active participants in Research and Innovation (R&I) projects, thus designing a new role for city authorities compared with previous framework programs where cities were only the place and the object where research took place.

Sustainable development of urban areas has become the prime challenge in the area of "Secure, clean and efficient energy", promoting transition to a competitive energy system around specific objectives such as energy consumption and carbon footprint reduction; low-cost and low-carbon electricity supply; a smart European electricity grid; alternative fuels and mobile energy sources; innovative knowledge and technologies; market uptake of energy and ICT innovation; robust decision making and public engagement.

Several funding schemes in the H2020 Framework Programme such as Smart Cities and Communities, European Structural and Investment Funds, European Fund for Strategic Investments, Urban Innovative Actions, Urbact, Life, and Jaspers have supported critical and ambitious urban dimensions.

As a consequence, the European Commission has been paying, for 30 years, significant attention to both urban dimensions and sustainable development, which are also cornerstones in its funding strategies; this trend seems to continue through the forthcoming Horizon Europe Framework Programme with a set of instruments and actions such as:

- Horizon Europe Cluster 5 [1]: Climate, Energy, and Mobility—Destination 2, Cross-sectorial solutions for the climate transition;
- European Partnership Driving Urban Transition to a sustainable future (DUT) partnerships [2];
- European research and innovation missions (European Green Deal) delivering solutions to some of the greatest challenges facing our world, among which is 100 climate-neutral cities by 2030—by and for the citizens [3].

In this broad picture, the SET Plan, which is the technology pillar of the EU's energy and climate policy adopted by the European Union, boosts the transition towards a climate-neutral energy system, promoting the development of low-carbon technologies in fast and cost-competitive ways. Thanks to new technology improvements and reducing costs with coordinated national research efforts, the SET Plan facilitates cooperation among EU countries, companies, and research institutions, delivering on the main challenges of the Energy Union.

The implementation of the SET Plan started with the establishment of the European Industrial Initiatives (EIIs) and of the European Energy Research Alliance (EERA), both promoting alignment of the Research and Development activities of individual research organizations to SET Plan challenges and creation of a joint programming framework at the EU level involving national delegates in Implementation Working Groups. Among the Implementation Working Groups, a specific one for Action 3.2 "Smart Cities and Communities" has been created, boosting planning, deployment, and replication of 100 Positive Energy Districts (PEDs) by 2025 for sustainable urbanization, thus anticipating the concept of PEDs highlighted in the European Partnership DUT.

On the same line, the EERA Joint Programme on Smart Cities (a network of universities and research institutes from across Europe that cooperate with industry, cities, and citizens to support innovation and demonstration projects) contributed to the development and mainstreaming of the Positive Energy Districts concept. The EERA Joint Programme on Smart Cities re-orientated its structure in 2018 to better investigate and progress the underlying PED concepts and to stimulate discussion around them.

With the aim to support capacity building for new generations of PED professionals, early career investigators, experienced practitioners, and policy makers, the EERA Joint Programme on Smart Cities has been leading the submission of a Cooperation in Science and Technology (COST) Action proposal to establish a European PED network. As a result, the Cooperation in Science and Technology (COST) Action CA19126 "PED-EU-NET Positive Energy Districts European Network" (Appendix B), supporting open collaboration among researchers, innovators, and other relevant stakeholders across different domains and sectors and giving impetus to research advancements and innovation on Positive Energy Districts in Europe—formally started its activities on 10 September 2020, under EU funding.

Cities with their surroundings areas have become the epicenter of the pandemic with ninety percent of all reported COVID-19 cases [4], exacerbating risks and taking extraordinary measures to support the local economy and protect the population, in some way losing the priority of persisting climate and ecological crisis [5]. Therefore, the COVID-19 pandemic clearly demonstrates interrelations between natural and societal systems; indeed, societal resilience strictly depends on a resilient and robust environmental support system [6].

As a consequence, with the outbreak of the COVID-19 crisis, emerging issues and related questions with respect to mainstreaming urban transition, sustainability, and climate neutrality arise spontaneously: How did societal challenges and related priorities change in the post-COVID scenario? Should the Urban Transition framework change as well? How much do innovative research and funding frameworks to boost transition already take those changes into account? How can addressing the challenges of climate neutrality help with respect to post-COVID 19 urban transition?

According to authors' point of view, tentative answers are provided in section "Discussion and Conclusions".

The COVID-19 crisis has had a deep economic impact on several sectors; indeed, the activity in construction decreased by 15.7% in 2019, as well as investment in the energy efficiency sector which decreased by 12% in 2020. Even if the European recovery plan is coming, a new impact on the building sector is expected which could be limited, somehow, by 2030, creating 160,000 green jobs for the EU construction sector thanks to the renovation wave. This would be a key opportunity to overcome the COVID-19 crisis, not forgetting about climate neutrality issues; restructuration and renovation offer a key opportunity to rethink, redesign, and modernize buildings to adapt them for a greener and digital society while sustaining economic recovery [7].

The following sections describe the most recent innovative research and funding frameworks—such as 100 Positive Energy Districts, 100 climate-neutral cities, Horizon Europe, and the Green Deal—aiming at offering critical perspectives taking into consideration the above dilemmas.

## 2. Materials and Method: A Review of the Most Recent Innovative Research and Funding Frameworks on 100 Positive Energy Districts and 100 Climate-Neutral Cities

More than half of the world's population nowadays is located in urban areas; people living in cities and their surrounding areas are expected to be 80% of population by 2050. Cities and urban areas are centers of economic activities, knowledge generation, innovation and new technologies, and places where the quality of life is immediately tangible. That is why Horizon Europe will promote the mission of "100 Climate-Neutral Cities by 2030—by and for the citizens" and will support sustainable urban transition with the DUT Partnership.

At the time of this article's writing, the definition of the contents relating to the 100 Climate-Neutral Cities Mission seems to have been fully outlined and defined, while the process of the DUT Partnership is still ongoing and it is assumed that it will not be completed before the second half of 2021. The process for understanding the aim, objectives, and priorities of the DUT Partnership is a co-creation process that has involved, and still involves, main urban stakeholders. Management of the DUT Partnership is entrusted to JPI Urban Europe, which, due to its role and skills, is also the management delegate for the Implementation Working Group of SET Plan 3.2.

In addition to many other urban stakeholders, two research networks participate in the co-creation process for DUT Partnership, mainly related to PEDs: the EERA Joint Program on Smart Cities and the COST Action CA19126 "PED-EU-NET".

Thanks to Horizon Europe and the co-design process, European cities will have a unique chance to promote the sustainability dimension at an urban scale, according to their own needs—this will help to fulfil the goals set out by international policy frameworks such as the Urban Agenda for the EU, the COP21 Paris Agreement, the Habitat III New Urban Agenda, and, last but not least, the UN's Sustainable Development Goals (notably SDG 11) of the DUT Partnership.

### 2.1. 100 Positive Energy Districts: Driving Urban Transition (DUT) Partnership

The aim of the European Partnerships in Horizon Europe is to address global challenges and modernize industry. The European Partnerships will contribute to multiple Sustainable Development Goals (SDGs) and are an integral part of Horizon Europe's strategic planning process that identifies three types of partnership: (a) Co-programmed European Partnerships, which are partnerships between the Commission and private and/or public partners; (b) co-funded European Partnerships using a program to co-fund action—this is the case for Driving Urban Transition (DUT) Partnership to a sustainable future; and (c) institutionalized European Partnerships with research and innovation funding programs that are undertaken by EU countries (e.g., the EIT KICs).

The Driving Urban Transition to a Sustainable Future (DUT) Partnership is the proposal for an urban partnership under Horizon Europe, led by JPI Urban Europe [8], and focuses specifically on fostering actions in Cluster 5 "Climate, Energy and Mobility", whose scope is to create impact in three urban dimensions: (a) urban areas, (b) urban policy, and (c) urban innovation, and the European Research Area (ERA).

The DUT Partnership addresses a complex set of urban challenges [9] and aims to support decision making in municipalities, companies, and society by acting and enabling:

1. Actions for capacity and community building for urban transformations to drive and to shape a quadruple-helix innovation eco-system on urban transitions. Targets: (a) City authorities and urban municipalities mobilized in projects, urban living labs, and other partnership activities; (b) Urban living labs and similar formats; (c) Multi-stakeholder engagement.
2. Efforts towards integrated approaches to tackle complex urban issues and to increase effectiveness of urban solutions, approaches, and processes. Targets: (a) R&I calls opened for participation of all stakeholder groups (challenge-driven); (b) Urban research and innovation considering urban dilemmas and engaging relevant stakeholders in the process.
3. Join forces—at local and international scales—to tap the full potential of urban Research and Innovation, bringing knowledge to action across Europe to create benefits for neighborhoods and urban areas across Europe. Targets: (a) Policy briefs and recommendations, with and made available for urban governance and city networks; (b) Widened participation and international relationships; (c) Results and solutions from R&I projects and/or urban living labs available for local urban policy and public administrations.

According to this approach, the Partnership will create a portfolio of measures to enhance its impact, build capacities in all stakeholder groups, and contribute to the European mission on climate-neutral and smart cities. On the other hand, the Partnership will engage and enable the whole spectrum of urban stakeholders (local authorities, municipalities, businesses, and citizens) to co-create innovative, systemic, and people-centric approaches, tools, methods, and services in support of urban transformative transitions. This will boost a more efficient and decarbonized use of energy, sustainable and people-friendly mobility systems, and circular and environmental-friendly use of resources for the well-being of citizens and preservation of biodiversity.

With this vision, the DUT Partnership is likely to contribute to global and European policies, in particular the Agenda 2030 and UN Habitat's New Urban Agenda, the strategic priorities of the European Commission's Strategic Plan for Horizon Europe, with a special focus on the European Green Deal, the Paris Agreement, the Leipzig Charter, and the Urban Agenda for the EU (UAEU). To achieve these objectives and to assess the sub-targets of the Sustainable Development Goal (SDG) no. 11 (Make cities inclusive, safe, resilient and sustainable), the DUT Partnership builds upon the Strategic Research and Innovation Agenda (SRIA) 2.0, developed by JPI Urban Europe, where three innovation pillars (and their inter-relationships) are considered as prioritized sectors along the Green Deal for sustainable urbanization:

1. The 15-Minute City as a concept for rethinking the urban mobility system and space;
2. Downsizing District Doughnuts as an integrated approach for urban greening and circularity transitions;
3. Positive Energy Districts and Neighborhoods, aiming at transforming the urban energy system.

#### 2.1.1. The 15-Minute City: Rethinking the Urban Mobility System and Space

The concept of the 15-Minute City (Table 1) provides a framework for the mobility of people and goods that directly and indirectly affects urban livability, health, the spatial configuration of cities, air quality, and other aspects of the living environment and sustainable urbanization. The transformation of urban mobility systems will not be achievable by relying on technologies or providing more sustainable mobility offerings alone but calls for fundamental rethinking of space and the re-organization of all daily activities.

**Table 1.** The 15-Minute City concept.

| 15-Minute City |
| --- |
| Rethinking urban mobility system and space:<br>- Fundamental rethinking of space and the re-organization of our daily activities<br>- Different mobility options at various scales (from neighborhoods up to regional scale)<br>- Different urban settings (city size, available mobility infrastructure, mobility supply/service patterns, etc.)<br>- Different mobility options at various scales (from neighborhoods up to regional scale)<br>- More flexible transport options along a broad set of innovations |

The 15-Minute City approach addresses different urban settings (city size, available mobility infrastructure, mobility supply/service patterns, etc.) and mobility options at various scales (from neighborhood up to regional scale), taking into account how people move and how goods and services are delivered, hence creating more flexible transportation options along a broad set of innovations—social, organizational, technological, and institutional. For these reasons, this concept offers a clear focus for the mobility pillar and a holistic, people-oriented, and challenge-driven perspective for distinct aspects of the doughnut economy and the overarching ambition of regenerative cities according a wider PED/PEN concept.

#### 2.1.2. The Downsizing District Doughnuts

Cities and urban areas are attractive starting points for making the global transition to livable societies and circular economies. The Downsizing Districts Doughnut (DDD) Pillar (Table 2) aims at operationalizing a dilemma-based approach to urban greening and circularity objectives of the Green Deal by supporting urban robustness through doughnut strategies in urban districts. The DDD approach holds a dual circles system concept of city doughnuts based on the exchanges/balance between an external ring—representing the nine "planetary boundaries"—and an inner ring including the city doughnuts' nine sectors. Between these two sets of boundaries, there is the ecologically safe and socially just space [10].

**Table 2.** The Downsizing District Doughnuts concept.

| Downsizing District Doughnuts (DDD) |
| --- |
| Doughnut strategies in urban districts:<br>- Dual circles system concept of city doughnuts<br>- External ring representing the nine "planetary boundaries"<br>- Inner ring including the city doughnuts' nine sectors |

According to this approach, the transitions driven by the Downsizing District Doughnuts paradigm can contribute to and promote a systemic change at all of the following four levels of urban transformation:

- Governance and mobilizing people for change, by supporting existing networks, groups, initiatives, and "urban doers" engaged in urban transformation activities.
- Circularity, by supporting transitions of metropolitan industries and new urban economies with circular growth.
- Sustainable spatial planning, by translating land-use and spatial/territorial planning into integrated urban development, by developing and improving Green-Blue Infrastructures (GBIs) and Nature-Based Solutions' (NBS) overall urban livability, public

health, and urban robustness together with cutting-edge approaches to clean-tech and entrepreneurial creativity.

- Urban design, by supporting the (re-)development of attractive built environments concerning livability and well-being as well as the urban design aspects of "high quality" public spaces.

#### 2.1.3. Positive Energy Districts

The PED concept, as said, implies an integrated approach for designing urban areas, districts, or groups of connected buildings producing net zero greenhouse gas emissions, managing an annual local/regional overflow production of renewable energy. The PED/PEN approach (Table 3) emphasizes the flexibility dimension of urban districts in the regional energy system based on renewable energy and addresses the ecological and energetic footprints of goods and services. Otherwise, PEDs include a proper consideration of user behavior and people's lifestyles without contradicting the guiding principles of cost efficiency, affordability of housing, and energy poverty.

**Table 3.** The PED concept.

| PED |
| --- |
| This approach emphasizes the flexibility dimension of urban districts in the (renewable) regional energy system and addresses the energetic and ecological footprint of goods and services:<br>- Energy efficiency<br>- Energy flexibility<br>- Energy production<br>- Quality of life, inclusiveness, sustainability, resilience and security of energy supply |

The PED Reference Framework [11] identifies three key functions for urban areas, related to context of their urban and regional energy system, that cities need to consider in their long-term urban strategies: energy efficiency, energy flexibility, and local/regional energy production. Therefore, the neighborhood scale will foster economic sustainability (e.g., economies of scale), aggregation synergies (e.g., efficiency deployment, flexibility, and integration), and governance in distributed resources through a considerable involvement of all stakeholders and communities. It means that each PED will design its own balance between the above-mentioned functions, figuring their own way towards climate neutrality and energy surplus. To make PEDs attractive for cities and citizens, their development should follow four guiding principles, namely: (a) quality of life, (b) inclusiveness (with special focus on the affordability and prevention of energy poverty), (c) sustainability, and (d) resilience and security of energy supply.

These three key priority areas—The 15-Minute City, The Downsizing District Doughnuts, and Positive Energy Districts—in turn, imply three integrated approaches that impact each other as well as most the other urban grand challenges, while revealing multiple interlinkages with the sectorial priorities [12]. The three DUT pillars can be considered complementary driven strategies (Tables 4–6) aiming at supporting climate neutrality and energy autonomy achievement in cities considering urban physical, functional, and socio-economic vulnerability. They create strong interfaces and cross-cutting potential towards each DUT pillar and suggest transversality, therefore overcoming a strictly silos vision ("breaking silos")—neighborhoods cannot produce more energy than they consume without new mobility solutions and more circular use of resources. New mobility solutions necessitate innovative energy technology and design thinking. Sustainable energy systems and smart mobility solutions are mandatory, should cities obtain circularity and secure well-being while respecting planetary boundaries. Otherwise, some commonalities across the three pillars suggest the renewal of the existing city by providing a new urban governance based on considering the expertise of residents/inhabitants. A change in the governance structures is required and can be improved by co-designing inclusive public spaces as different ways for people to interact with technology and nature.

**Table 4.** Interface of the 15-Minute City with the other two Driving Urban Transition (DUT) pillars.

| PED | Downsizing District Doughnuts (DDDs) |
| --- | --- |
| Developments on neighborhood scale.<br>- Energy demands for mobility<br>- Infrastructure requirements<br>- Functional design and planning of neighborhoods | Contribution to greening of urban areas and economies.<br>- Impact on material flows and circularity<br>- Mixed use of neighborhoods and link to green-blue infrastructure<br>- Shared economy |

**Table 5.** Interface of the Downsizing District Doughnuts with the other two DUT pillars.

| Positive Energy Districts | 15-Minute City |
| --- | --- |
| The role of energy in doughnut districts:<br>- Refurbishment and upgrade of built environment<br>- Energy and resource efficiency and circularity of PEDs<br>- Integration of GBIs and NBS in PEDs<br>- Energy solutions for new urban manufacturing | Interplay of the 15-minute city and doughnut district:<br>- Contribution of new mobility systems and services to circular urban economies<br>- Mobility solutions for metropolitan industries<br>- Consideration of the urban-rural continuum<br>- Multi-functional public spaces |

**Table 6.** Interface of Positive Energy Districts with the other two DUT pillars.

| Downsizing District Doughnuts | 15-Minute City |
| --- | --- |
| PED contribution to DDD:<br>- Integration of green-blue infrastructure in PED concepts<br>- Impact of PED on circularity, material flows, and resource efficiency<br>- Economic dimension of PEDs | Mobility in PEDs:<br>- Integrated energy and mobility planning<br>- Developing attractive, inclusive, and accessible neighborhoods or districts<br>- Strengthen the role of mobility in neighborhood concepts |

The PED concept itself represents the result of several working groups and ongoing initiatives at the European level that involve members from the EERA, JPSC, and JPI UE network. Even if an accurate concept of PEDs is still under discussion, an extensive range of different approaches and actions at the European and international levels are currently focusing on the definition and the implementation of the PED strategy for energy transition and climate mitigation in the urban context, among which are:

- The International Energy Agency EBC Annex 83 "PEDs", launched in February 2020 and focused on developing an in-depth definition of PEDs and the technologies, the planning tools, and the decision-making processes related to Positive Energy Districts;
- The already mentioned COST Action CA19126 "PED-EU-NET" (Positive Energy Districts European Network) supporting open collaboration among researchers, innovators, and other relevant stakeholders across different domains and sectors to drive the deployment of Positive Energy Districts in Europe;
- The Positive Energy District Booklet which contains more than 60 examples of PEDs existing, ongoing, or in transition, showcasing different aims and challenges, approaches, and factors of success.

### 2.2. 100 Climate-Neutral Cities by 2030: Innovation Mission European Green Deal

The European Commission has recently introduced its own strategic long-term vision for a prosperous, competitive, and climate-neutral economy by 2050. The "European Green Deal" makes the case for a mission-oriented R&I component in the Horizon Europe program to give answer to the question of how Europe can help cities become climate-neutral as fast as possible; supporting, promoting, and showcasing hundreds of European cities in their structural change towards climate neutrality by 2030, transforming urban areas and districts into hubs of experimentation and innovation and being able to lead on the European Green Deal and on Europe's efforts to be climate neutral by 2050.

To achieve in ten years within European cities what Europe plans to achieve in 30 years is a huge challenge that requires a systemic transformation for acting on the global climate emergency and for delivering co-benefits that will improve the health, well-being, and prosperity of citizens. This transformation will be feasible because technologies and innovative solutions for sustainable energy, transport, food, water, and material systems already exist thanks to R&I programs of last decade and in the years to come due to Horizon Europe and national R&I programs.

Moreover, green technology prices and market conditions are fast moving towards climate-friendly investments and will continue to strengthen incentives to transition. The European Green Deal, including a revision of EU directives for 2030, and the new role of the European Investment Bank will, thus, further strengthen this trend.

The mission (Table 7) is based on two main pillars:

- The Multiannual Financial Framework MFF 2021–2027 and Next Generation EU proposals which will impact the European and national frameworks for the funding of climate action, including the Horizon Europe program where the mission is anchored in terms of objectives, R&I agenda, and societal challenges and priorities;
- The European Green Deal which sets an unprecedented level of ambition and reach for climate and environmental action and for the financing and inclusiveness of the transition.

**Table 7.** The Climate-Neutral Cities concept.

| Climate Neutral Cities |
|---|
| To set new standards for climate and urban agendas and for their implementation, addressing the challenge of climate neutrality:<br>- Cities will act as innovation hubs and national, European, and global forerunners<br>- Climate City Contract<br>- Climate neutrality should explain the starting point in its societal, economic, ecological, and political dimensions<br>- Connection with the local/regional or national strategy for carbon neutrality by 2050<br>- The "cross-border" issue: the participating city should ensure that measures taken will not be physically unconnected or stop working at the borders of the selected site |

The innovation Mission Green Deal is designed as a flagship initiative that complements and blends the two mentioned pillars, aiming to set new standards for climate and urban agendas and for their implementation. While connected to a wide range of European policies and strategies (e.g., the Climate Law and the Climate Pact, the EU plan for circular economy, the European long-term strategy for 2050 and the National Energy and Climate Plans, the Urban Agenda for the EU, the European Digital Strategy, the Smart Specialization Strategies and platform, and the Smart and Sustainable Mobility Strategy), the mission will also connect these policies and strategies at the local level. The participating cities will act as innovation hubs and national, European, and global forerunners. They will inspire additional urban areas, and eventually, the whole of Europe, to accelerate their policies for climate action and transition.

To address the challenge of climate neutrality, the mission proposes a multi-level co-creation process through the introduction of a Climate City Contract. The purpose is to (a) express the ambition and commitment of all involved parties to the mission objectives; (b) identify the policy and implementation gaps as a basis for a strategy for transition; (c) coordinate stakeholders and empower citizens in the city around a common climate goal; (d) coordinate national/regional and EU authorities to deliver the necessary legal, governance, and financial framework conditions to support each city; (e) create a one-stop shop for multi-level negotiations to facilitate city action for transition.

Adapted to the specific circumstances of each city, a Climate City Contract will include the goal and targets, specify the strategy and the action plan for transition, and identify stakeholders and responsibilities.

The Contract is not meant as a closed document that only binds a city legally to a course of action up to 2030. Instead, it emphasizes the high ambition, the participatory approaches, and the multi-level governance collaboration that will trigger innovation and change towards climate neutrality.

Participating cities will be encouraged to design and implement a multi-sector governance model where local stakeholders such as business, universities and the civil society are part of the Climate City Contract and contribute to its design and implementation in and for the city (Table 8). Similarly, regional or national stakeholders that can ensure the success of the Contract (e.g., regional transport companies, national energy producers, national research institutions etc.) should be encouraged to join by the relevant regional or national signatory.

**Table 8.** Interface of Climate-Neutral Cities with two DUT pillars.

| PED | 15-Minute City |
| --- | --- |
| CNC contributions to PED:<br>- Connection with local/regional dimensions<br>- Impact not only on areas physically connected<br>- Multi-sector governance model | CNC contribution to 15-Minute City:<br>- Identify the policy and implementation gaps<br>- Coordinate stakeholders and empower citizens |

Thanks to the Innovation Mission Green Deal, the current system for funding and financing climate innovation and investment at city level—now too fragmented—will be innovated; the commitment by the European Commission to a Sustainable Europe Investment Plan offers a new ambitious framework with a strengthened role for the European Investment Bank (EIB). Furthermore, the new EU Multiannual Financial Framework (MFF), reinforced by the recovery plan, will be an enabler and accelerator of the needed shift, combining financing from different European, national, and local resources and the whole value chain from Research and Innovation to planning, investment, and implementation.

When ready for investment, cities will be able to apply for a variety of instruments/funds/facilities, mainly involving EIB funds. In fact, the estimated urban lending by EIB for 2012–2018 is EUR 152 billion, out of which nearly EUR 26 billion were invested in climate mitigation actions which include Natural Capital Financing Facility and Municipal Loans. Other instruments, i.e., Connecting Europe Facility, can provide guarantees and bonds, whereas the European Energy Efficiency Fund (a PPP with international banks) can provide cities with market-based junior debt, guarantees, equity, mezzanine instruments, leasing structures, and forfeiting loans.

## 3. Discussion and Conclusions

A strong global European leadership in renewable energy technologies coupled with circularity, resilience, and sustainability will pave the way to increase energy security and reliability in times of crisis, as can be seen today during the COVID-19 crisis, and beyond [13]. It will also lead to achieving the objectives of the European Green Deal as well as to sustain economic recovery and growth in the long term while ensuring a sustainable future for European citizens. Both the EU long-term climate strategy and the European Green Deal highlight the importance of renewable energies for a future clean European Energy System with achievement of the zero-emission target. Although a broad range of technologies for climate neutrality are already expected, European Commission programming for Research and Innovation should also leave space for those emerging and break-through technologies with a high potential to achieve climate neutrality. These technologies will have a significant role in reaching the EU's goal to become climate-neutral by 2050.

Nevertheless, the challenge is not just technical—it calls for wide-ranging societal transformations and the adaptation of lifestyles and behaviors, even more so after the experienced COVID-19 pandemic, also to cope with the expected economic emergency during the next few years [14].

The COVID-19 crisis has highlighted the significance of homes and buildings for our lives as well as their congenital fragilities. In the pandemic period, the home has been the core of daily life for Europeans, an office for those teleworking, a nursery or classroom for children, and a hub for online shopping or downloading entertainment. During the pandemic, buildings with a proper destination were empty as schools had to start experimenting with distance learning, private business companies had to rearrange to social distance, and public offices had one side implementing smart working and the other side managing social distancing to provide public service. Last but not least, hospital infrastructures have been under hard pressure. Some effects of the pandemic will continue in the longer term, stimulating growth for new demands on existing buildings and their energy and resource profiles, adding the need to deeply renovate European building stocks.

Emphasis must be given to place-based approaches and experimentation capitalizing on citizens' engagement, social innovation, and user-led and citizen science for shared ownership of solutions that are tailored to local specificities and thus ensure that envisaged transition pathways are human-centered and just. Important gaps in knowledge, evidence, innovation, technology, data, capacity, and skills, lack of integrated approaches, and deficit in applying research and innovation results to actions exist and prevent successful implementation of such transitions. Furthermore, institutional fragmentation and non-inclusive and non-participatory governance structures lead to a lack of shared vision, goals, and direction regarding the transition process, incoherence in policies and strategies, uncoordinated planning and decision-making, ineffective measures, and inefficient use of resources.

Climate neutrality and good environmental conditions, accessibility to qualitative urban open space, and a more inclusive urban economy appear even more necessary in a post-COVID-19 scenario [15]. Recent research has highlighted that "the specificity of an increase in the number of cases of contagion that affected some areas of Northern Italy in particular could be linked to the conditions of pollution by atmospheric particulate matter which carried out a carrier and boost action" [16].

The COVID-19 crisis drove attention to the importance of improving cities' resilience attitude; the pandemic highlighted not only the connection between air quality and well-being to health crises but also between food/medicine freight transport and survival, between health services and citizens, and between employment and living in cities; the COVID-19 crisis also emphasized the importance of buildings and the urban shell which support our lives and defend our fragilities.

What emerged after the past nine months and is new to come is that it seems that citizens, while continuing to assess as very important the actions necessary to protect the environment, ask more for an economic and employment recovery, which appears to be vital for everyone.

It seems equally necessary to clearly focus on sustainability in the urban dimension as a factor of economic growth, as the need to renovate cities and their dynamics can offer a unique opportunity to redesign our cities and their social services, urban planning, public transport, health services, and urban food logistics, integrating sustainability strategies from the early beginning.

Our reaction to the COVID-19 crisis, which somehow brings our mind back to the 2008–2009 recession, can now benefit from ten years of technological development, which could facilitate a joint goal of economic recovery and environmental and health protection, responding to the now impellent choice for sustainability.

The choice for sustainable development, only a few months ago, was seen by scientists, politicians, the world of finance, and business leaders as the only possible one. Now, due to the COVID-19 crisis, the transition towards a sustainable and more affordable urban model is not a possibility but a compulsory target, as the EU already included resilience as a priority in all urban long-term strategies.

The Horizon Europe Framework Programme—thanks to Destination 2 Cross-sectoral solutions for the climate transition, Communities and Cities; Positive Energy Districts paradigm; DUT Partnership, and Innovation Mission 100 Climate-Neutral Cities—will boost the transition towards a climate-neutral energy system in cities promoting the quadruple-helix innovation eco-system for urban transitions [16,17]. This goal appears, in principle, to be possible thanks to the innovative system for funding climate innovation and investment at the city level.

Now that technologies and innovative solutions for sustainable energy, transport, food, water, and material systems already exist, thanks to Research and Innovation programs developed over the last decade, what is left to tackle the climate-neutrality challenge is to have relevant stakeholders from academia, industry, cities, and communities to actively participate in the open innovation process transcending disciplinary, administrative, hierarchical borders and, mostly, municipalities able to address these "wicked dilemmas" [18].

Among the key concepts highlighted in the previous sections such as the re-organization of our daily activities and mobility supply/service patterns (15-minute cities), the resilience and security of energy supply (PEDs), and, last but not least, climate neutrality (Climate-Neutral Cities) will certainly be at the core of the forthcoming Horizon Europe Programme, as well as among the top priorities for cities and communities alike.

**Author Contributions:** Conceptualization, P.C.M., M.B.A. and P.C.; methodology, P.C.M., M.B.A. and P.C.; formal analysis, P.C.M., M.B.A. and P.C.; investigation, P.C.M., M.B.A. and P.C.; writing—original draft preparation, P.C.M., M.B.A. and P.C.; writing—review and editing, P.C.M., M.B.A. and P.C.; visualization, P.C.; supervision, P.C.M. and M.B.A.; All authors have read and agreed to the published version of the manuscript.

**Funding:** Paolo Civiero who is one of the co-author of present paper has received funding from the European Union's Horizon 2020 research and innovation programme under the Marie Skłodowska-Curie grant agreement No. 712949 (TECNIOspring PLUS) and from the Agency for Business Competitiveness of the Government of Catalonia. TECNIOspring PLUS investigator: Paolo Civiero, Project: PEDRERA. Positive Energy Districts renovation model. IREC—Catalonia Institute for Energy Research (Barcelona—ES).

**Conflicts of Interest:** The authors declare no conflict of interest.

## Appendix A. Framework Programs 5, 6, and 7

The Fifth Framework Programme (FP5) focused on a problem-solving approach thanks to "key actions", among which was one on "The City of Tomorrow and Cultural Heritage" as a first integrated and systemic R&I approach to urban challenges. The City of Tomorrow and Cultural Heritage Key Action aimed to improve urban sustainability through delivering benefits to citizens by concentrating resources on specific areas of city planning and management, cultural heritage, built environment, and urban transport. The Key Action was specifically designed to ensure rapid, EU-wide take-up of practical new approaches to urban governance, planning, and management. Total FP5 EC Contribution to Urban Research projects 1998–2002: circa EUR 479.0 million.

The Sixth Framework Programme (FP6) introduced funding for smart cities and supported the European Research Area and the ERA-NETs with EU top-up funding to joint calls and other coordination activities among national programs. The Joint Programming Initiative Urban Europe (JPI UE) is the output of such an ERA-NET on urban issues. Total FP6 EC Contribution to Urban Research projects 2002–2006: circa EUR 400.3 million.

The Seventh Framework Programme (FP7) deepened comprehension on urban challenges and has been fundamental for the development of technology as well as the development of strategies for sustainable urban development in different areas, notably energy efficiency and climate action. Total FP7 EC Contribution to Urban Research projects 2007–2013: circa EUR 1.9 billion.

## Appendix B. COST Action CA19126 "Positive Energy Districts European Network—PED-EU-NET"

In the last decade, building in Europe has significantly supported the level of innovation as seen in the development of Nearly Zero-Energy Buildings (NZEBs). It is now time to step up efforts towards city-wide transformation with the pioneering concept of Positive Energy Districts (PEDs), which builds on the paradigm of smart cities.

A Positive Energy District has been defined as "energy-efficient and energy-flexible urban areas or groups of connected buildings which produce net zero greenhouse gas emissions and actively manage an annual local or regional surplus production of renewable energy. They require integration of different systems and infrastructures and interaction between buildings, the users and the regional energy, mobility and ICT systems, while securing the energy supply and a good life for all in line with social, economic and environmental sustainability" [9].

This represents a major challenge that crosses sectors and domains whose solutions would only be found through collective innovation. "Innovation" implies opening up the relevant processes to all active players so that knowledge can flow across the entire economic and social environment [19]. The deployment of 100 Positive Energy Districts and Neighborhoods and 100 Climate-Neutral Cities requires innovation in multiple domains, encompassing interconnected technological, social, cultural, political, spatial, economic, and regulatory aspects. Each domain has its own set of embedded challenges that need to be addressed in order to foster the innovation process. In terms of technological challenges, innovative concepts, products, and services are needed to produce optimal PED/Climate-Neutral City solutions customized to local circumstances. Moreover, companies in the building, energy, mobility, and ICT sectors need to develop new business models for the emerging PED/Climate-Neutral Cities market. They need to draw on flows of knowledge outside their boundaries to boost the internal innovative processes [20]. In the social aspect, the processes of societal innovation, social entrepreneurship, and citizen participation must be integrated synergistically in the transformation [21]. In the financial aspect, robust investments and creative funding models are needed to support innovative energy solutions and establish new energy markets. With respect to legal aspects, regulatory sandboxes are needed to test novel solutions on PED/Climate-Neutral Cities. Requirements for the certification and standardization of PED/Climate-Neutral Cities need to be defined to ensure quality and facilitate replication.

The concept of PEDs is evolving and it still needs to be refined, advanced, demonstrated, implemented, and replicated [22]. Europe is poised to enable transitions towards a climate-neutral economy and the concept of PEDs will be incrementally introduced in the energy planning of many cities and communities in the coming years [23]. According to the SET Plan Action 3.2, 100 PEDs are expected, as said, "to be in concrete planning, construction or operation, synergistically connected to the energy system in Europe by 2025" [24].

Within this framework, the main aims and objectives of the COST Action CA19126 "PED-EU-NET" have been set to "drive the deployment of PEDs in Europe by harmonizing, sharing and disseminating knowledge and breakthroughs on PEDs across different stakeholders (academia, industry, cities and communities), domains (technological, social, economic, financial, legal and regulatory) and sectors (buildings, energy, mobility and ICT) at the national and European level. [The PED-EU-Network aims at supporting] cities and empower communities to achieve city-wide positive energy transformation with pioneering ideas, methods and solutions. It will mobilize the relevant actors from and across Europe to collectively contribute to the long-term climate neutral goal" [25].

This COST Action is structured into four interlinked Working Groups: three are scientific Working Groups focusing on a variety of topics related to PEDs (WG 1, 2, and 3) and a dissemination Working Group (WG 4) that oversees the communication, education, outreach, and exploitation of this Action's outcomes.

The COST Action CA19126 PED-EU-NET builds on a holistic approach, bringing together stakeholders from different backgrounds. This initiative represents a strong commitment to pooling resources, experimenting with new methods, co-creating original solutions, and advancing science and building solutions so as to enable neighborhoods and cities to integrate different functions and guiding principles and find the balance that represents the best renewable energy resources usable in their respective climate zones, according to their communities' specific ambitions and needs [26].

## References

1. Verso Horizon Europe. Available online: https://www.versohorizoneurope.it/ (accessed on 3 November 2020).
2. Presentation of the Driving Urban Transitions to a Sustainable Future (DUT) Partnership. Available online: jpi-urbaneurope.eu (accessed on 3 November 2020).
3. DG RTD (EC). *100 Climate-Neutral Cities by 2030—By and for the Citizens*; Publications Office of the European Union: Luxembourg, 2020; ISBN 978-92-76-19919-9. [CrossRef]
4. United Nations Human Settlements Programme (UN-Habitat). Opinion: COVID-19 Demonstrates Urgent Need for Cities to Prepare for Pandemics. 15 June 2020. Available online: https://unhabitat.org/opinion-covid-19-demonstrates-urgent-need-for-cities-to-prepare-for-pandemics (accessed on 30 November 2020).
5. Saadat, S.; Rawtani, D.; Hussain, C.M. Environmental perspective of COVID-19. *Sci. Total Environ.* **2020**, *728*, 138870. [CrossRef] [PubMed]
6. Sharifi, A.; Khavarian-Garmsir, A.R. The COVID-19 pandemic: Impacts on cities and major lessons for urban planning, design, and management. *Sci. Total Environ.* **2020**, *749*, 142391. [CrossRef] [PubMed]
7. European Commission. A Renovation Wave for Europe—Greening Our Buildings, Creating Jobs, Improving Lives, COM (2020) 662. Available online: https://eur-lex.europa.eu/legal-content/EN/TXT/?qid=1603122220757&uri=CELEX:52020DC0662 (accessed on 30 November 2020).
8. JPI Urban Europe. Available online: https://jpi-urbaneurope.eu/governance/governing-board/ (accessed on 3 November 2020).
9. JPI Urban Europe. Driving Urban Transitions to a Sustainable Future. Proposal for an European Partnerships under Horizon Europe. Report (version July 2020). Available online: https://ec.europa.eu/info/files/european-partnership-driving-urban-transitions-sustainable-future-dut_en (accessed on 3 November 2020).
10. Raworth, K. A Doughnut for the Anthropocene: Humanity's compass in the 21st century. *Lancet Planet. Health* **2017**, *1*, e48–e49. [CrossRef]
11. JPI Urban Europe. PED Programme Management. Europe towards Positive Energy Districts. A compilation of Projects towards Sustainable Urbanization and the Energy Transition. Booklet (Version February 2020). Available online: https://jpi-urbaneurope.eu/app/uploads/2020/02/PED-Booklet-Update-Feb2020.pdf (accessed on 3 November 2020).
12. JPI Urban Europe. Strategic Research and Innovation Agenda 2.0. Available online: https://jpi-urbaneurope.eu/app/uploads/2019/02/SRIA2.0.pdf (accessed on 3 November 2020).
13. IEA. The Impact of the Covid-19 Crisis on Clean Energy Progress, IEA, Paris. 2020. Available online: https://www.iea.org/articles/the-impact-of-the-covid-19-crisis-on-clean-energy-progress (accessed on 3 November 2020).
14. Kunzmann, K.R. Smart Cities After Covid-19: Ten Narratives. *Disp-Plan. Rev.* **2020**, *56*, 20–31. [CrossRef]
15. Italian Society of Environmental Medicine (SIMA). Position Paper Particulate Matter and COVID-19. 16 March 2020. Available online: http://www.simaonlus.it/wpsima/wp-content/uploads/2020/03/COVID_19_position-paper_ENG.pdf (accessed on 30 November 2020).
16. Campbell, D.F.J.; Carayannis, E.G.; Rehman, S.S. Quadruple helix structures of quality of democracy in innovation systems: The USA, OECD countries, and EU member countries in global comparison. *J. Knowl. Econ.* **2015**, *6*, 467–493. [CrossRef]
17. Carayannis, E.G.; Barth, T.D.; Campbell, D.F.J. The quintuple helix innovation model: Global warming as a challenge and driver for innovation. *J. Innov. Entrep.* **2012**, *1*, 1–12. [CrossRef]
18. Wrangsten, C.; Bylund, J. A Dilemma-Driven Approach to Urban Innovation. JPI UE. 2018 (Online Version). Available online: https://jpi-urbaneurope.eu/news/a-dilemma-driven-approach-to-urban-innovation/ (accessed on 30 November 2020).
19. EC. *Open Innovation, Open Science, Open to the World—A Vision for Europe*; Publications Office of the European Union: Luxembourg, 2015.
20. Errichiello, L.; Marasco, A. Open service innovation in smart cities: A framework for exploring innovation networks in the development of new city services. *Adv. Eng. Forum* **2014**, *11*, 115–124. [CrossRef]
21. Haarstad, H.; Wathne, M. Are smart city projects catalyzing urban energy sustainability? *Energy Policy* **2019**, *129*, 918–925. [CrossRef]

22. Aghamolaei, R.; Shamsi, M.H.; Tahsildoost, M.; O'Donnell, M. Review of district-scale energy performance analysis: Outlooks towards holistic urban frameworks. *Sustain. Cities Soc.* **2018**, *41*, 252–264. [CrossRef]
23. Koutra, S.; Becue, V.; Gallas, M.A. Towards the development of a net-zero energy district evaluation approach: A review of sustainable approaches and assessment tools. *Sustain. Cities Soc.* **2018**, *39*, 784–800. [CrossRef]
24. SET-Plan ACTION No. 3.2. Implementation Plan. (Version June 2018). Available online: https://setis.ec.europa.eu/system/files/setplan_smartcities_implementationplan.pdf (accessed on 3 November 2020).
25. COST Action CA19126. Memorandum of Understanding. 2020. Available online: https://www.cost.eu/actions/CA19126/#tabs\T1\textbar{}Name:overview (accessed on 3 November 2020).
26. JPI Urban Europe. Positive Energy Districts (PED). 2020. Available online: https://jpi-urbaneurope.eu/ped/ (accessed on 3 November 2020).

MDPI
St. Alban-Anlage 66
4052 Basel
Switzerland
Tel. +41 61 683 77 34
Fax +41 61 302 89 18
www.mdpi.com

*Energies* Editorial Office
E-mail: energies@mdpi.com
www.mdpi.com/journal/energies

www.ingramcontent.com/pod-product-compliance
Lightning Source LLC
LaVergne TN
LVHW070918170726
843515LV00005B/812

* 9 7 8 3 0 3 6 5 1 1 8 8 7 *